University-Industry Collaboration and the Success Mechanism of Collaboration: Case Studies in Japan

The front cover was designed based on an image originally designed by Georgios Tsorman in Greece

RIVER PUBLISHERS SERIES IN INNOVATION AND CHANGE IN EDUCATION - CROSS-CULTURAL PERSPECTIVE

Volume 8

Nowadays, educational institutions are being challenged when professional competences and expertise become progressively more complex. This is mainly because problems are more technology-bounded, unstable and ill-defined with the involvement of various integrated issues. To solve these problems, it requires interdisciplinary knowledge, collaboration skills, innovative thinking among other competences. In order to facilitate students with the competences expected in professions, educational institutions worldwide are implementing innovations and changes in many aspects.

This book series includes a list of research projects that document innovation and change in education. The topics range from organizational change, curriculum design and innovation, pedagogy development, to the role of teaching staff in the change process, students' performance in the aspects of not only academic scores, but also learning processes and skills development such as problem solving creativity, communication, and quality issues, among others. An inter- or cross-cultural perspective is studied in this book series that includes three layers. First, research contexts in these books include different countries/regions with various educational traditions, systems and societal backgrounds in a global context. Second, the impact of professional and institutional cultures such as language, engineering, medicine and health, and teachers' education are also taken into consideration in these research projects. Thirdly, individual beliefs, perceptions, identity development and skills development in the learning processes, and inter-personal interaction and communication within the cultural contexts in the first two layers.

We strongly encourage you as an expert within this field to contribute with your research and make an international awareness of this scientific subject.

For a list of other books in this series, www.riverpublishers.com
http://www.riverpublishers.com/ series.php?msg=Innovation_and_Change_in_Education_-_Cross-cultural_Perspective

University-Industry Collaboration and the Success Mechanism of Collaboration: Case Studies in Japan

Zhiying Nian

Beijing Institute Learning Society (BILS)

River Publishers

Published, sold and distributed by:
River Publishers
Niels Jernes Vej 10
9220 Aalborg Ø
Denmark

River Publishers
Lange Geer 44
2611 PW Delft
The Netherlands

Tel.: +45369953197
www.riverpublishers.com

ISBN: 978-87-93379-04-6 (Hardback)
978-87-93379-03-9 (Ebook)

Contents

Foreword I

The higher education serves a threefold function: teaching, research, and social service. It is generally assumed that the function of teaching for the cultivation of talent has been a cornerstone of the university since the Middle Ages; the function of research was developed at Martin Luther University, Halle-Wittenbergand Georg-August-University of Göttingen, and especially Humboldt-Universität zu Berlin in Germany; while social service became a significant function in the wake of the advent of land-grant universities in the late nineteenth century. Universities in East Asia, however, especially research universities, did not develop a social service role for a long time.

One of the main causes of this is the minimized impact of the American land-grant universities on East Asia. As the flagship university in the modern time, Peking University was developed into a research university due to President Yuanpei Cai's preference for pursuing profound theoretical knowledge as opposed to the applied knowledge. Furthermore, several subjects based on applied knowledge were removed from the university curriculum by President Cai himself. Before the middle of twentieth century, universities organized by American clergymen did have a great and broad impact on the character of liberal education from Britain, while land-grant universities were seldom found at that time. In Japan, the first Minister of MEXT (the Ministry of Education, Culture, Sports, Science and Technology), Arinori Mori, introduced the German university model implementing the first change to Japan's university system since the Meiji Restoration. Thus, the land-grant universities only influenced a few of the higher education institutions in Japan, such as Sapporo Agricultural College. Pursuing pure scientific research at a distance from the society has been seen as the core ethos of universities, especially high-level universities, for a very long time.

But this attitude towards the function of the university changed in the late twentieth century in Japan. As a rare-successful industrial country in Asia, Japan implemented the Strategic Orientation of Developing Country via Science and Technology development strategy, which focused on technology innovation, including a wide variety of world-class technologies in the fields of auto industry, electronics, chemistry, energy, etc., which promoted the rapid growth of the economy. Enterprise, which has little to do with universities, is the main driver of technological innovation; nevertheless, the strong pursuit of technology ultimately connects enterprises and universities.

Academic research was seen as the highest virtue in the research universities of Japan. Meanwhile, systems that seldom encouraged, and even restricted, professors from participating in research studies run by enterprises had little benefit for the universities. Influenced deeply by the successful American cases of university-industry collaboration, the Japanese government decided in the 1980s to foster the contact between universities and enterprise through a series of science and technology policies (e.g., the TLO Act). In the middle of the 1990s, Japan issued a Basic Law on Science and Technology and announced the slogan of Developing Country via Science and Technology Innovation, which promoted the policy environment of university-industry collaboration.

This significant change in Japanese universities was admired by China. As a matter of fact, many Chinese universities and colleges had participated in the research and development of enterprises before the Reforming and Opening-up Policy. Most of the universities that participated in these programs were led by official industry administrations such as railway, oil, textile, geology, steel, etc., in order to be allowed to collaborate with industry, providing the corresponding technology support for enterprises. But the collaborations mentioned above fell under orders coming from government administrations, without the context of a market economy. When the Ministry of Education began regulating these universities, they could not maintain their close ties with those industries, which led to a new, emerging problem: how to realize their social service function. Meanwhile, most of the comprehensive universities were baffled on how to establish university-industry collaboration because they mainly focus on basic research, without worrying about the economy.

Dr. Zhiying Nian started her study under these circumstances, with financial support provided by the China Scholarship Council and with the help of Professor Yutaka Otsuka of Hiroshima University to conduct the fieldwork in Japan for one year. Unlike other Chinese scholars focusing on the study of Japanese government and its drafted policies, she concentrated on investigating the collaborations between universities and enterprises that seem to be unrelated to each other at the micro-level in practice. For the sake of this research, Dr. Zhiying Nian visited the key people responsible for these projects, who had rich experience in collaboration, and their students participated in the project and the relevant project management offices at the University of Tokyo, Waseda University, and Hiroshima University, even approaching their enterprise counterparts. After interviewing 45 people, she ultimately chose 20 samples from among them, leaving her with an almost 28-h transcript. With this hard work, she created a fundamental understanding of university-industry collaboration in Japan, which is quite different from the research findings of government reports. Her efforts to involve herself in fieldwork were highly regarded by her dissertation defense committee.

Without doubt, this study still has some small defects from the perspective of internal mechanisms and institutional environment. This is a result of the fact that most of the information from the projects that the universities and enterprises collaborate on are need to be confidential, and that the author has had to construct her entire understanding of collaboration from fragmented and dodged explanations, so that she can hardly give a precise accounting of the degree of motivation in the collaboration between the two parties. Dr. Zhiying Nian's efforts to explore the real-world practice of university-industry collaboration have, therefore, been invaluable.

It is expected that this book will not only serve as the significant base for the author's research career, but will also open a window of opportunity for university-industry collaboration in China.

Institute of International and Comparative Education,
Beijing Normal University,
Yimin Gao, Ph.D.
Aug. 31st, 2014

Foreword II

The author of this book, Dr. Zhiying Nian, was awarded a scholarship under the State Scholarship Fund of China to pursue her study in Japan as a joint Ph.D. student. Since her academic advisor in China, Professor Yimin Gao of Beijing Normal University, is one of my old friends, and asked me to be her advisor in Japan, I was very delighted to introduce her to Hiroshima University, where I worked until retiring at the end of March, 2014. Zhiying, calling her as usual by her first name, stayed in Hiroshima for 1 year, from April 2012 to March 2013. During this period, she worked very diligently and with great perseverance to conduct the fieldwork and collect related information to complete her dissertation under the title "University-Industry Collaboration of Japanese Higher Education."

At Hiroshima University, Zhiying was deeply involved in the work at its Center for the Collaborative Research and Community Cooperation, which functions to promote industry–academia–government–community relations. Professors and staff of the Center kindly gave her advice, as well as chances to participate in various seminars and workshops related to her research theme. She did not stay exclusively on our campus, but extended her fieldwork to some Japanese companies, including Mazda and Nissan, as well as some other universities, including the University of Tokyo and Waseda University, as typical examples of "university-industry collaboration" in Japan. Due to her very social and friendly character, almost everybody applauded her research pursuits. After collecting a huge volume of information, she tried to build-up an analytical framework and devoted all her energy to writing a dissertation draft. During the process, Zhiying and I discussed the framework and structure of the dissertation. Although I was her advisor, as an old Chinese proverb says, both teacher and student made progress by learning from each other. I am confident that I have really benefited from the experience of being her advisor, and my days with her left me many precious memories.

While in Hiroshima, in addition to devoting a lot of time to her own research work, Zhiying also attended one of the seminar courses on comparative education that I was teaching. When she attended my classes, I was deeply impressed by her attitude: she was earnestly involved in the class every week. She often took the initiative and led the discussion in the class. Such an attitude exerted a beneficial effect on other students. Based on her fieldwork and various experiences in Japan, Zhiying not only finished and submitted her dissertation to the Beijing Normal University immediately after returning home, and successfully passed the defense, but also had a chance to publish it in a short period of time. In Japan, more and more young scholars who have just graduated from their doctoral course wish to publish their dissertation immediately, but the opportunity to realize their wish is so limited. I think the same is true also in China. The fact that Zhiying's dissertation will be published soon indicates its excellence in quality. I would like to express my hearty congratulations on this. I feel very proud to write this preface for Dr. Zhiying Nian's forthcoming book and also firmly believe that the book will be welcomed and appreciated by many readers.

Yutaka Otsuka, Ph.D.
Professor, Fukuyama University
Professor Emeritus, Hiroshima University
Former President, Japan Comparative Education Society

List of Figures

List of Tables

List of Abbreviations

AGS	Alliance for Global Sustainability
AIST	(National Institute of) Advanced Industrial Science and Technology
CARB	California Air Resources Board
CASTI	Center for Advanced Science and Technology Incubation, Ltd.
DUCR	Director of the Division of University Corporate Relations
ETF	European Training Foundation
HTC	Hiroshima Technology Transfer Center
IMD	International Institute for Management Development
IP	Intellectual Property
JARI	Japanese Automobile Research Institute
JAXA	Japan Aerospace Exploration Agency
JST	Japanese Science and Technology Agency
MEXT	Ministry of Education, Culture, Sports, Science and Technology
MITI	Ministry of International Trade and Industry
MLITT	Ministry of Land, Infrastructure, Transport and Tourism
NPOs	Non-Profit Organizations
NTSEL	National Traffic Safety and Environment Laboratory
PROs	Public Research Organizations
R&D	Research and Development
RCAST	Research Center for Advanced Science and Technology
RIETI	Research Institute of Economy, Trade and Industry
SIR	Senior Innovation Researcher
TLF	Technology Liaison Fellow
TLO Act	Act on Promotion of Technology Transfer from Universities to Private Business
TLOs	Technology Licensing Organizations
TOUDAI	University of Tokyo

UIC	University-Industry Collaboration
UIGC	University-Industry-Government Collaboration
UIRC	University-Industry-Research Institute Collaboration
UNESCO	United Nations Educational, Scientific and Cultural Organization
UNISPAR	University-Industry-Science Partnership
UNITT	University Network for Innovation and Technology Transfer
USPTO	US Patent and Trademark Office
UTC	United Technologies Corporation
UTEC	University of Tokyo Edge Capital Co., Ltd.
VBL	Venture-Business Laboratory
WTLO	Waseda TLO

1

Introduction

1.1 The Rising Role of the University in the Knowledge-based Society

In the new era of continuous development of the knowledge-based economy, in which the value of university-industry collaboration (UIC) gains greater recognition, UIC consequently becomes a more significant subject of study in the higher education. There is no doubt that this is a result of UIC boosting the development of economy and society. With the accelerated global economic integration, knowledge innovation becomes a decisive factor for countries looking to enhance and maintain their competitiveness. However, UIC not only emphasizes knowledge creation, but also highlights the significance of knowledge transfer and application to a large extent. In the 1980s, Freeman, a British scholar, introduced the concept of "National Innovation System" in *Technology and Economic Performance: Lessons from Japan*, a research monograph on Japanese economic development. He believed that the National Innovation System was "the network of institutions in the public and private sectors, activities and interactions, which help initiate, introduce, change, and diffuse new technologies" (Freeman, 1987, p. 76). In 1997, OECD further underlined the significance of the interaction among various social actors in its publication *National Innovation Systems*:

> Innovation is thus the result of a complex interaction between various actors and institutions. Technological innovation is a result of interaction and feedback among all elements within the system rather than an occurrence in a perfect linear fashion. The core of this system is enterprises and the way they organize production, innovation and channels by which they gain access to external sources of knowledge. Main sources of external knowledge are other enterprises, public or private research institutions, universities and intermediary organizations (OECD, 1997).

This implies, enterprises, research institutions, universities, and intermediary organizations constitute the main part of the National Innovation System through collaboration and interaction. For the first time, the Chinese Academy of Sciences formally introduced the National Innovation System in *Greeting the Era of the Knowledge-based Economy and Construction of the National Innovation System*, a research report submitted to the Central Committee of the Communist Party of China at the end of 1997. The *Outline of the National Medium- and Long-Term Program for Science and Technology Development (2006–2020)* also clearly suggested "building an enterprise-led and market-directed technology innovation system featuring the combination of enterprises, universities, and research institutions, which will be a breakthrough point for the full-fledged construction of the national innovation." In light of the policy guarantee and support from the Chinese government, UIC has been established as the central strategy for the construction of the National Innovation System, and will furthermore be promoted and developed as a long-term strategy.

On the other hand, the huge effect that UIC has on universities has also come into the public eye. John L. Hennessy, the President of Stanford University, once said:

> It is said that there would have been no Silicon Valley without Stanford University, but I want to add that there would have been no first-class Stanford University without Silicon Valley. The biggest advantage of locating Stanford University in Silicon Valley is that we can get what the enterprises do, estimate what kind of problems they may encounter and then solve for them in advance. Science and technology parks can help the university perform its education responsibilities. The second advantage is that inviting elites from industrial circles to give classes may enable students to know more about the world and the society as well as their future work environment. And, the teachers, if wanting being involved in industrial matters, do not need to walk a long way. Although we are living in a digital era, short distance is still very important and beneficial for face-to-face communication. It is also convenient for personnel in industrial circles to come to the university. Both the university and enterprises appreciate the respective roles that they play (Luo & Li, 2009).

Evidence from the most profitable university patents in America indicated that universities' interest in basic research, knowledge spreading, and resource demand did not always hinder the collaborative participation of professors and their assistants in entrepreneurial inventions (Branscomb, Florida, & Kodama, 2003). After a 20-year period of promotion and development of UIC, the University of Warwick in Britain ranked among the top British universities. On that account, the United Nations Educational, Scientific and Cultural Organization (UNESCO) issued the *World Declaration on Higher Education* in Paris on October 9, 1998. Article 4 of this declaration states that the effectiveness of higher education should be assessed based on the degree of agreement between what society expects from colleges and universities and what they actually do. For this reason, colleges and universities should particularly focus on strengthening the communication with industrial circles, as well as on establishing a long-term development orientation based on the societal needs.

Despite its great significance, UIC is not a simple undertaking. For example, in China, UIC is confronted with many serious problems, such as insufficient motivation and an unpolished system. Specifically, problems are as follows: they have a low-cooperation level due to a non-uniform risk allocation; insufficient cooperation depth due to the absence of a long-term cooperation mechanism; frustration or even termination of the collaboration on the project due to an unclear definition of responsibility for the capital input in university-industry research institute collaboration (UIRC); cooperation lag due to incompatibility of the goals from industry and university, etc. (Gong, Ge, & Chen, 2009; Zhang & Wu, 2008; Tan, 2010; Zheng & Liu, 2004; Zhou, Yuan, Gu, & Li, 2010). These problems exist not only in China, but also in many other countries in varying degrees.

Many causes lead to the suspension or failure of collaboration projects. First, universities and enterprises theoretically have different laws of development. Although both have the same objective of enhancing the development of society and the economy, they have specific and distinct responsibilities. Smooth collaboration between universities and enterprises can only be realized through an exploratory process of trial and error, rather than being done at one go. Second, UIC may not be actually necessary due to the restrictions in economic development levels, industrial structures, etc. UIC is truly required only when knowledge plays a crucial role in the economic development; when the enterprises cannot survive without intellectual support, especially basic research support; and when the importance of universities participating in technology transfer becomes more pressing. Once this kind of demand for UIC has been generated, it requires a certain economic foundation as an internal guarantee of smooth cooperation. Finally, UIC is also restricted by management systems, including economic and social systems. UIC cannot continue without the external support of these systems. In addition to the above mentioned causes, more complex reasons also lead to the suspension or failure of projects. In the pressing search for UIC, exploring practical UIC experiences from different countries and institutions, and trying to make them work for our reference serves as an important channel for discovering the dynamics of UIC. In recent years, the academic circle in China has emphasized the study of UIC in foreign countries. Specifically, it has focused on America, where UIC is comparatively mature; Japan, where the government plays an active role in UIC; and Europe and other regions, where UIC also plays a role.

Although the relative weakening of Japanese comprehensive economic strength in the recent years has attracted much attention domestically, Japan was still ranked No. 1 for its innovation power by the World Economic Forum in the *Global Competitiveness Report 2012–2013* issued in 2012. Japan's position can be mainly attributed to the research and development funds from enterprises, the advantages of these enterprises' overseas markets, and the implementation of new technologies by these enterprises, and further attributed to the quality of Japanese scientific research institutions and to the efficacy of collaborative research and development under UIC (Table 1.1). Thus, research on UIC in Japan is of very obvious significance to China. However, academia in China has mainly focused on the macro system of UIC in Japan based on the policy text of the government, exploring in particular the government's role in UIC, while neglecting the specifics of its practice. Precisely due to the fact that Chinese researchers are nowhere near Japan's level of collaborative research between universities and enterprises, their attempts to research the UIC policies abroad amount to less effective solutions when applied in practice in China.

In view of the reasons laid out above, the author believes that incorporating the Japanese practice of UIC into the literature on UIC is of both theoretical and practical significance. Theoretically, a deep understanding of the specific practice of UIC can lead us to the development of a theory of the internal "laws" of UIC. With this foundation, we can elucidate the specific motivation and obstruction of UIC, the different roles of universities and enterprises in promoting it, and factors affecting the success of UIC. At a practical level, given a deep understanding of the specific practice of UIC in Japan, we can logically explain political intention and the construction of the system, as well as finding out the specific roles played by universities and enterprises in UIC, in order to sort out what best practices may be helpful for universities in China. Furthermore, research universities play a leading role in knowledge

Table 1.1 Global competitiveness index (2012) rankings

Country	Capacity for Innovation	UIC in Research and Development (R&D)	R&D Funds from Enterprises	Quality of Scientific Research Institutions	Innovative and High Technologies Developed	Innovative Technologies Implemented by Enterprises	Advantages of Overseas Markets of Enterprises	Technologies Introduced by Overseas Investment
Japan	1	16	2	11	11	4	2	67
China	23	35	24	44	107	71	56	77
USA	7	3	7	6	14	14	18	43
UK	12	2	12	3	6	23	6	35
Switzerland	2	1	1	2	2	3	1	36

Source: Prepared based on the data from Klaus Schwab. "The World Economic Forum (2012)", *The Global Competitiveness Report of 2012–2013*.

innovation. Their collaboration with enterprises has accelerated the construction of the National Innovation System. Therefore, the author has selected three research universities (the University of Tokyo, Waseda University, and Hiroshima University) as subjects of investigation, and will analyze their practice of UIC in order to summarize UIC's basic features.

The questions on specific issues are addressed in this study and it includes: (i) what is the main motivation to enhance UIC in Japan? (ii) What institutional structures guarantee the process of UIC? (iii) What are the different reciprocal requirements of universities and enterprises? (iv) What are the positive effects of UIC on universities and enterprises? (v) What constitutes the main obstruction to UIC? (vi) What are the requirements for the self-improvement of universities and enterprises in UIC? (vii) What are the fundamental contradictions arising from UIC?

1.2 The Development of UIC

The concepts of "UIC," "UIRC," and university-industry-government collaboration "(UIGC)" point in the same fundamental direction, although they are slightly different. It would seem too narrow to represent the two fields of "knowledge creation" and "knowledge application" with only two words, "enterprise" ("industry") and "university" ("university," including a wide variety of other higher education institutions), respectively, but the collaboration subjects that "industry" and "university" represent is relatively clear on the whole. This book will not investigate the role of scientific research institutions, and only focus on the relevant system arranged by the government that will be considered as the external conditions for UIC, so "UIRC" and "UIGC" will not be used. In other words, this paper only applies the concept of "UIC," and interprets it as enterprises and universities carrying out a series of activities in the form of technical collaboration, personnel communication, consulting, achievements transfer, establishment of new enterprises, etc., based on their own comparative advantages, with the support of the government, scientific research institutions, financial organizations, etc. UIC is an important part of the National Innovation System.

This book focuses on a study of the practice of university-industry collaboration in the research universities in Japan. The author has fully researched the material covered using domestic and overseas academic resources from the National Library of China, China National Knowledge Infrastructure (CNKI), WorldCat Dissertations and Theses, EBSCO Academic Search Premier, Calis Current Contents of Western Journals, Springerlink, etc.

Studies on UIC in China largely began in the 1990s, with *The New Stage for Higher Education Development—Relationship between University and Industry*, written by Hui Xu (a professor at Zhejiang University), being particularly groundbreaking and influential. Professor Chengxu Wang, a famous expert on comparative higher education, wrote in his preface that this book introduced current areas of study on UIC, and also summarized the abundant existing experience in developing university-industry relationships and attendant problems. He furthermore gave his incisive opinions on university-industry relationships, combining theory and practice by systematically investigating and analyzing them based on typical information from the historical development and current situations of these relationships in various countries (Xu, 1990, p. 12).

A literature review of journals in CNKI from 1979 to March 12, 2011 shows that relevant studies on UIC were published in core journals at a high frequency, including in *Research in Education Development, China Higher Education Evaluation, China Higher Education Research, Research on Higher Education of Engineering* and *Heilongjiang Research on Higher Education,* before 2000. After 2000, however, these studies were mainly published in the academic journals of various higher vocational colleges, not in core journals. According to this analysis, the studies of UIC carried out by higher vocational colleges and by higher education account for 42.5 and 11.2% of all journal articles, respectively, and those on UIC in foreign countries account for only 9%. Therefore, the study of UIC in vocational colleges is more active than the study of UIC in other institutional settings, but focuses primarily on practice and the improvement of collaborative modes. UIC was gradually implemented at the end of the 1990s in some universities of higher education; examples include studies on the practice of UIC at Tsinghua University (Electronics and Packaging, 2010), Peking University (Gu, 2007), Shanghai University of Engineering Science (Chen, 2008, Beijing Institute of Technology), East China University of Science and Technology (Guo, 1999), Donghua University (Gao et al., 1999), Tongji University (Gu et al., 1997), etc. With the constant expansion of UIC in higher education as a whole, most scholars have focused on the motivation behind the practice of UIC in this field since 2010. It can therefore, be concluded that the literature on UIC in higher education as a whole has been reduced due to the bottleneck encountered during the construction of an effective platform for collaboration between universities and enterprises. It, therefore, becomes necessary to create an effective institutional environment as a motivation to promote the bilateral collaboration and attain the goals of knowledge innovation, talent cultivation, quality increase in universities, and scientific and technological progress in enterprises.

1.2.1 Development of UIC in the World

The modes of UIRC are distinguished based on specific situations. The knowledge and resources that are exchanged by collaborating parties determine these modes, which are also influenced by actual collaborative targets and experiences. Furthermore, the significance of these impact factors may change with time.

Shantha Liyanage, an early UIC researcher, summarized these UIRC modes as collaborative enterprises, direct contracts, technology licenses, manufacturing or marketing agreements, exchange and transfer of researchers, Cohen, Nelson, and Walsh (2002), as well as other researchers, evaluated the flow path of useful information from universities and other public research institutions to the industrial research with the Carnegie Mellon Survey. The survey evaluated the modes generally used in UIRC with a four-point Likert scale. The study indicated that publications and written reports were the main path in UIC, followed by informal information exchange, public meetings, and consulting. The ratio of the companies achieving UIC through hiring graduates who participated in UIC projects, or the companies jointly establishing enterprises, applying for patents, etc. was significantly lower than the abovementioned modes. Technology licenses and short-term individual exchanges were believed to be the least frequently used modes on the path to UIC.

Annamária Inzelt (2004) classified UIC according to the degree of embodiment of two collaborating parties in her research. Among the 18 collaboration modes listed in Table 1.2, modes 1–5 are in the scope of exchange between two collaborating parties, while modes 6–16 are within the scope of collaboration, with mode 15 realizing research and development collaboration based on contracts, and mode 16 realizing formal collaboration through joint establishment of enterprises. Mode 16 is the most complicated collaborative mode with the highest degree of collaboration in UIRC.

Ina Drejer and other researchers analyzed the modes of UIC in Denmark through research and development centers, joint cultivation of doctors, and joint inventions produced in their research (refer to Table 1.3). They found that the effect of high-level knowledge production in universities and research institutions lies in the "overflow" process in industry, not in a direct increase of industrial technical compatibility. Such "overflow" is generally

Table 1.2 University-industry-government collaboration and its classifications

Mode	Collaboration Level	Diagram of Exchange Mode
1. Specific training for employees of enterprises at universities 2. Lectures given to employees at universities 3. Training for university teachers in enterprises 4. Informal exchange between people at universities and those in enterprises through professional league meetings, forums, and other forms 5. Purchasing of research outcomes (patents) from universities	Between individuals	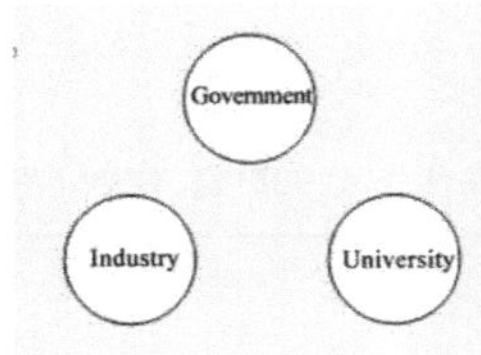
6. Employing university professors as long-term consultants 7. Mentoring of employees of enterprises by researchers from universities (cross-organization mentoring relationship) 8. Training of employees of enterprises by university professors 9. Joint publications by professors from universities and employees of enterprises 10. Jointly cultivating doctors and masters students by universities and enterprises 11. Sharing of IP by university professors with employees of enterprises	Between individuals Between organizations	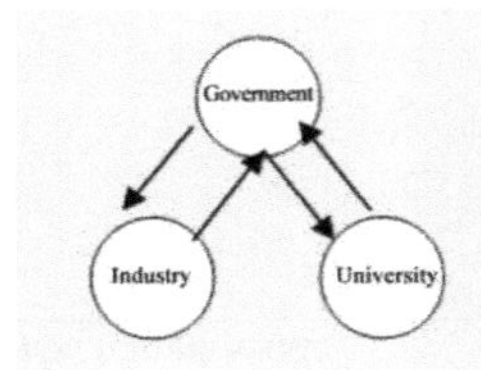

(*Continued*)

Table 1.2 Continued

Mode	Collaboration Level	Diagram of Exchange Mode
12. Ability of universities or enterprises to use their counterpart's equipment without assistance 13. Investing in equipment for universities by enterprises 14. Enterprises regularly obtaining research outcomes from universities 15. Contract research—formal research and development collaboration 16. Collaborative research projects—formal research and development collaboration 17. Establishment of permanent knowledge flow from universities to enterprises 18. Producing knowledge flow through overflow effect from joint establishment of enterprises	Between individuals	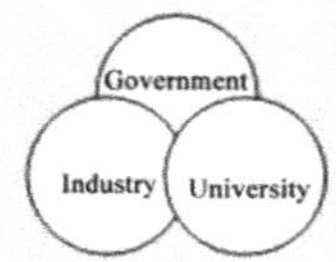

Source: Annamária Inzelt. "The evolution of university-industry-government relationships during transition". *Research Policy*, 33 (2004): 975–995.

Table 1.3 Modes of university-industry-research institute collaboration and contents

Mode	Effects
Contract-based research center	In this mode, a research center is established by joint contributions from one or more universities and enterprises, and at least one authorized technical service institution. The collaboration foundation is research projects with commercial value.
Industrial doctor	Industrial doctors is a program designed for enterprise employees in collaboration with universities and other research institutions aimed at constructing a network of research institutions and enterprises to promote the development of economy, technology, and other aspects of the whole network.
Invention law	The law is issued to manage the salaries of inventors at public research institutions and enable better productization of their research achievements.

Source: Drejer, I. and Jørgensen, B. H. "The dynamic creation of knowledge: Analyzing public–private collaboration". *Technovation*, 25 (2005): 83–94.

realized by achievement transfer to enterprise application from teaching and administrative staff members and students engaged in research in universities. Their research also showed that the collaboration between the industrial circle and academic research units actually created technology and knowledge innovation, even if university-industry collaboration was less than between academia and enterprises, and that low frequency of UIC was due to the lack of institutional support.

1.2.2 Development of UIC in Japan

The practice of UIC in Japan became widespread in the 1990s, and relevant academic research began accordingly. Prominent Japanese scholars engaged in studying UIC include Kazuyuki Motohashi (元桥一之), an economics professor at the University of Tokyo; Professor Yasunori Baba (马场靖宪), from the Research Center for Advanced Science and Technology (RCAST) of the University of Tokyo; and Masahiko Aoki, an honorary professor at Stanford University.

Low (1997) indicated that the Japanese Ministry of Education had specially formulated four collaboration modes, including collaborative research, contract research, transfer of university technology, and enterprise innovation, by taking advantage of the research achievements of universities in order to stimulate and strengthen the collaboration between enterprises and academia, since 1983. Yoshiro Sawada (1990, 2003), a Japanese scholar, pointed out that the relationship between enterprises and universities clearly showed hierarchical models, which mainly involved the single-layer model in individual relations; the two-layer model in the industrial and academic circles and government sectors, as well as in the systems in which collaborative research was considered; and the three-layer model, which incorporated both individuals/organizations and systems. In recent years, due to the changes in strategic alliances for research, development activities of Japanese enterprises, transitions in operation modes in universities, and societal demand for the creation of new industries and jobs, UIGC has attracted wide attention from the public, and scholars have gradually began to discuss the effects and mechanisms of UIGC at both national and regional levels. Etzkowitz, Leydesdorff (2000)[1], and Harayama (2003) analyzed the contributions of universities in the National Innovation System to the economic development and the modes of collaboration between universities and industries. In addition, the Japanese scholars Mitsui (2004) and Ito[2] (2000) studied the role of universities in regional economic development using regional innovation system theory. In 2003, the Ministry of Education, Culture, Sports, Science and Technology (MEXT) specified the following five typical UIGC modes:

- For research: collaborative research by enterprises and universities and contract research;
- For education: practice from universities in enterprises and joint development of educational programs;
- For technology licensing organizations (TLOs): transfer of scientific achievements of universities to enterprises by way of technology licensing;
- For consulting: consulting activities by researchers based on part-time technical instruction;
- For entrepreneurship: entrepreneurial activities based on research achievements and human capital potential of universities.

For years, the UIC system implemented by MEXT consisted of collaborative research, contract research, contract researchers, contribution funds for lectures on an academic subject, contributed lectures, and other modes through which collaborative research and contract research could obtain direct support.

[1]Etzkowitz, H. and Leydesdorff, L. "The dynamics of innovation: from national systems and 'mode 2' to a triple helix of university-industry-government relations". *Research Policy,* 29 (2000): 109–123.

[2]Ito Masao (伊藤昭男). "Sangakurenkei and Regional Innovation (産学連携と地域イノベーション)". *Kitami University Paper Collection* (北見大学論集). 23 (2000): 13–35.

With Masayuki Kondo's (2004) publication, he became the authority on the modes of UIGC in Japan in the context of knowledge production, transfer, and entrepreneurship, and divided them into three modes: joint knowledge creation, knowledge transfer, and knowledge-based entrepreneurship. The mode of joint knowledge creation consists of collaborative research, contract research, and scholarship. The mode of knowledge transfer consists of patent exchange, technical training, technical negotiation, technical consulting, and employment of researchers. The mode of knowledge-based entrepreneurship consists of venture businesses in universities and entrepreneurial universities (Figure 1.1).

UIGC in Japan follows a distinctive model. After deciding to promote economic development in the country using innovation in science and technology, Japan began its long-term exploration into practice. Large enterprises and national universities in Japan-dominated innovation, and the practice of UIC in Japan featured a structured division of labor: universities mainly conducted basic research; research institutes that mainly conducted applied research; and enterprises that mainly conducted development research. In order to adapt to the challenges of globalization and informatization in the twenty-first century, however, new approaches arose in UIC in Japan, and as a result this ordered division of labor began to breakdown. The Research

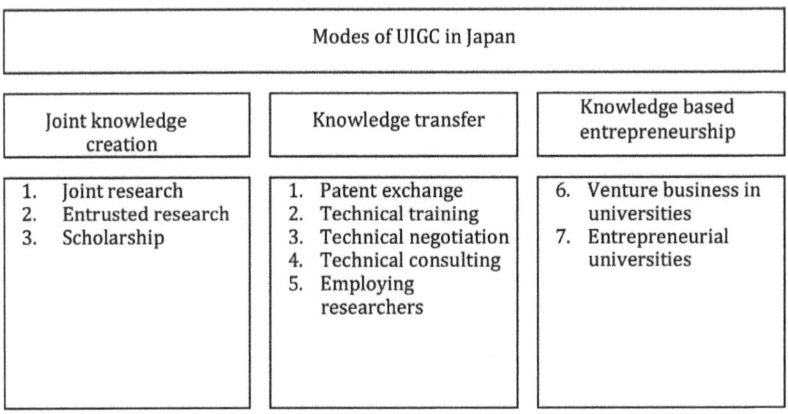

Figure 1.1 Typical modes of UIGC in Japan.

Source: Masayuki Kondo (近藤正幸), "Reform in Science and Technology Policy in Japan": *Science and Technology Policy to Innovation Policy*, 19(2004): 132–140. Edited by the author according to Masayuki's ideas.

Institute of Economy, Trade and Industry (RIETI) conducted a "survey on the current situation of collaborative research between enterprises and external institutions" and found that about 70% of the surveyed enterprises were carrying out collaborative research and development in various forms with external institutions (Motohashi, 2003, p. 4). These collaborative partners include large enterprises, small and medium-sized enterprises, venture capital enterprises, universities, etc. The number of projects done collaboratively as a proportion of all projects had increased when compared to 5 years ago (Figure 1.2), and many enterprises planned to increase it further.

In relationship between research and development under UIC, there are two approaches for enterprises: enterprises generally believed that independent research and development meant *turn[ing] research achievements into commodities faster* or *try[ing] to develop the core technology*; while research and development under UIC mean researching and developing new products for enterprises, particularly researching and developing the *latest technology* and producing the *basic research* carried out under UIC.

Kazuyuki Motohashi (2003) conducted an empirical quantitative analysis of UIC in Japan and found that the modes of UIGC were closely related

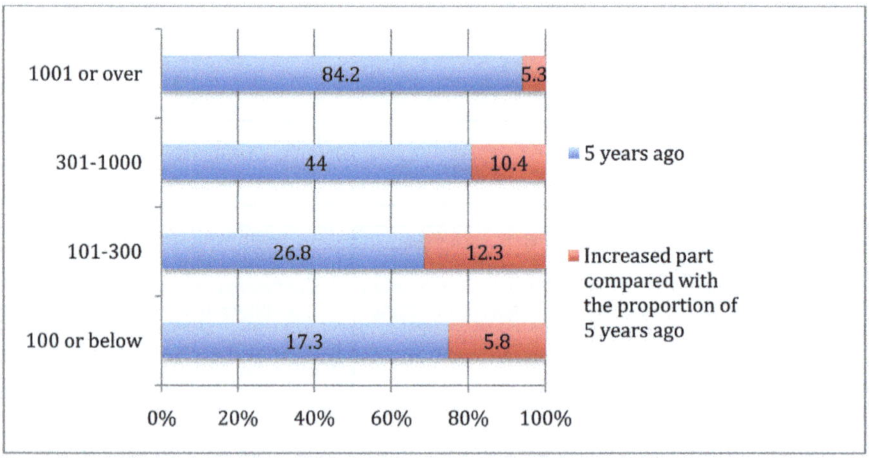

Figure 1.2 Current implementation of UIC based on number of enterprise staff (Unit: %, 2002).

Source: Kazuyuki Motohashi (元桥一之). "The quantitative analysis on university-industry collaboration practice and effect" 2003 (http://www.rieti.go.jp/jp/publications/dp/03j015.pdf).

to the scales of enterprises and universities, and that collaborative research accounted for the highest proportion of UIC. In large enterprises with more than 1,000 staff members, collaborative research accounted for 80% of the modes of collaboration with universities, followed by contract research and scholarship. In addition, in small and medium-sized enterprises, collaborative research between universities and the industrial circle accounted for about 50% of collaboration, and technical negotiation and consulting accounted for a relatively higher proportion compared with other approaches. The following presents the reasons for engaging in UIGC. Most enterprises expected to "increase the basic research level," "obtain the basic research achievements of collaborative partners," and so on through this kind of collaboration. As for the time expected for research and development results to turn into commodities, based on different collaboration objectives, 80% of the collaboration within enterprises was expected to be completed "within 1 year" or to last "2–3 years." Furthermore, about half of the enterprises involved in UIGC answered that the duration of the collaboration was "5 years or longer," and applied different collaborative modes to different requirements of research and development.

In 2005, the National Institute of Science and Technology Policy of Japan published a report on UIGC and regional innovation development that specifically analyzed which of the main modes of UIGC increased regional innovation capacity in Japan, and included a contrastive analysis of the main modes of collaborative innovation between enterprises and universities (in the current development system of collaborative research and contract research).

In terms of universities, collaborative research and contract research in collaboration with enterprises and national universities dominated the modes of UIGC when compared with public or private universities. Since the issue of the Law of Science and Technology in 1995, and the implementation of the first and second Science and Technology Basic Plans in 1996, the amount of collaborative and contract research has naturally increased. Since 2002, the amount of collaborative research has exceeded that of the contract research in innovation of UIGC due to the implementation of the Law of Industrial Technology Reinforcement, the Intellectual Property Law, the State IPR Strategy, the second Science and Technology Basic Plan, the corporatization of national universities, and the implementation of other systems and policies. The confidence in collaborative organizations has consequently increased greatly in the process of innovation from UIGC. There are two reasons for this: the expansion and deepening of UIGC demanded by the internal requirements for globalization and informationization, and the promotion of various policies and measures by the Japanese government.

Through comparison of different UIC modes, a number of scholars have come to the conclusion that those driven by enterprises have incomparable advantages. Ning Ma (2006) indicated that, in practice, enterprises must have a leading position in UIRC, which will realize the constant technological innovation within the market system and fully play the role of UIRC in technology innovation; this was also the basic conclusion of a new economic reform. Lang Xu and Dong Liu (2006) also argued that the impact of enterprises is inestimable in UIRC, and took a leading position by conducting a general survey of UIRC development history. Yiming Wang and Jun Wang (2005) underlined the notion that the function of government in strengthening enterprises' capability for independent innovation was to create a new environment that favored enterprise innovation, strengthened effective protection for intellectual property (IP), enhanced breakthroughs in technical chains with broad schemes and plans, increased input and supply for common technologies, and promoted enterprise-dominated UIRC.

1.2.3 The Policy and Environment of UIC

The number of university patents granted has increased greatly after the Bayh-Dole Act was passed by the United States' in 1980, and consequently many studies on university patents emerged. The Bayh-Dole Act operated under the assumption that significant technologies generated by universities were not developed or applied effectively, and that the application of inventions (of which the universities keep ownership) in commercial fields could promote the development of the industry. However, although the number of university patents increased after the passage of the act, their quality was scarcely discussed.

From 1965 to 1992, a total of 19,535 university patents were registered by the US Patent and Trademark Office (USPTO). Henderson (1998) selected 1% of the patents registered at the USPTO during this period in order to evaluate changes in quality, relevance, and universality of university patents. Generally, university patents demonstrated higher significance and universality than common patents; but the significant statistical difference between the two rates only lasted until 1982. Indeed, with the rapid increase in the number of university patents issued, their significance and universality decreased. Some have argued that universities lacked patent experience, or decrease in the scientific research quality of universities; Henderson's results indicated that the Bayh-Dole Act and other policies motivated universities to apply for patents and simultaneously encouraged them to apply for insignificant inventions.

Some studies reported contradictory conclusions. Sampat (2003) and other scholars selected the same samples chosen by Henderson in his study in 1998, and combined the data from 1992 and 1999. But they found that the indicator of significance did not decrease, concluding that there were data truncation problems in Henderson's original samples. University patents have a longer quotation delay than enterprise patents, meaning that the data provided must cover a longer span of time in order to avoid statistical inaccuracy created by a lower number of patents registered in some years compared to others. However, some studies indicated that the significance of patents depended on the selection of data cited. Lanjouw (2004) and some other scholars believe the data quoted earlier was more valuable, while others, including Hall (2005), prioritized the more recent data.

In another re-evaluation, Mowery (2002) divided the universities evaluated by Henderson (1998) into three classifications: (i) universities with ten or more patents applied for before the Bayh-Dole Act was passed, (ii) universities with fewer than ten patents applied for before the Bayh-Dole Act was passed, and (iii) universities that applied for patents only after the Bayh-Dole Act was passed. The results showed that only the value of patents in classifications (ii) and (iii) exhibited a downward trend. Mowery and Henderson believed that the quality of patents from first-class universities did not decrease, but the rapid influx of universities without patent experience into the patent application pool led to a decrease in the overall quality of university patents after the Bayh-Dole Act was passed. Furthermore, after the act was passed, the number of university patents increased substantially, but began to fall again after a high point around 2000, which further strengthened scholars' objection to the significance of university patents (Leydesdorff, 2010).

Did the Bayh-Dole Act really affect the nature of scientific research in universities? Did the intense motivation of the commoditization of scientific output result in ignorance of the basic research conducted by universities? Rossel and Elara surveyed the quotation focus in university inventions and found a decrease in the quality of university patent knowledge. Finally, they found that the statistical interest rate of university patent citation was significantly lower than that of enterprise patent quotation, but that gap had been gradually closing since 1983. The downward trend in university knowledge production was noticeable mainly in pharmaceutical and biological fields, and no obvious difference could be found in other subjects. Furthermore, based on the results of the questionnaire, Thursby (2002) argued that universities had created a way to radically pursue the industrialization of university technology using more enterprise characteristics.

American UIC modes not only spread to Japan, but they were also very popular in Europe. Mowery and Sampat (2005), and Thursby (2002) argued that the Bayh-Dole Act only slightly affected university patents in America, and indicate that the number of patents applied for by universities in America had been increasing before the issue of the act. Mowery and Thursby also argued that American universities traditionally kept a close relationship with NGO organizations and enterprises, which could have a potential impact on other countries. Even if similar policies were implemented in other countries, their effects might be limited due to differences in environment (for instance, university traditions). Moreover, they asserted, no matter what kind of policies and measures that the Bayh-Dole Act has inspired the OECD countries (Denmark, Germany, France, Canada, and Japan) to take, similar policies based on American modes would ignore the core premise of the act (i.e., that the publicly sponsored research results were meant to transfer from the national level to the university level and to focus on the inventions of researchers in universities).

Goldfarb and Henrekson (2003) compared the industrialization policies for university IP in America and those in Sweden, and found that both countries provided very high budgets for university R&D, but applied different transfer modes for research achievements. By analyzing the technology transfer mechanisms of university IP, they concluded that the reason for the failure of university IP industrialization in Sweden was the inefficacy of the university incentive structure. Although technology transfer required the active participation of inventors in universities, neither the technology transfer in the universities nor from the universities were included in the scope of academic performance. In the United States, various policies, including the Bayh-Dole Act, provided commercial motivation to universities, generally moving decisions from the bottom to the top of the ladder, so that universities were able to choose the optimum scheme through constant trial and error. Goldfarb and Henrekson also indicated that the IP belonged to universities and not to researchers, thereby increasing the researchers' motivation to industrialize technology. TLOs provided effective support for patent applications, IP licensing, and other activities. In Swedish modes, they insisted that the government formulate IP policies on behalf of universities, but gave universities little decision-making power, as in other European countries. American modes were based on competition between university researchers and universities, while

Swedish modes expanded the power of universities but did not change the overall system of the universities, so that Swedish modes were questioned.

These studies focused on the different forms of organization in the universities of the United States, Europe, Japan, etc. and their corresponding institutional environments. They did not provide systematic analysis of patent data. For instance, Henderson (1998) and Sampat (2003) studied the cases in America. Tamada and Eno (2007) examined the change in the number of patents applied for by collaborations between universities and public research organizations (PROs) working with enterprises, and the results showed that the number of UIC patents had risen steadily since 1972 and then taken a sharp increase at the end of the 1990s. Kanama (2008) and other scholars completed a detailed analysis of the fields in which patents were applied by the University of Tsukuba, Hiroshima University and Tohoku University. Although the IP activities of these three universities were different in scale and covered a variety of fields, they had something in common: the number of patents applied in collaboration had increased, while the number of enterprise patents applied for by university researchers as inventors had decreased to some extent. Baba and Goto (2007) studied changes in UIC activities in universities. They conducted a survey at the School of Engineering and the School of Biological Sciences at the University of Tokyo to investigate the ways institutional reform promoted UIC activities by tracking UIC during the implementation of reform, from 1998 to 2003. (715 questionnaires were distributed and 402 were returned.) Through investigating the collaborative research conducted by researchers at universities with those at other organizations, they found that 85% of collaborative research was conducted with other university researchers and that 79% of collaborative research was conducted with large national enterprises. The proportion of collaborative research with small and medium-sized enterprises (SMEs) increased from 30 to 51% in 2003. Their results also showed that at least 60% of the researchers applied for one or more patents while 20% of them or more signed licensing agreements, and that informal conventional collaboration practice was sustainable and extensive. While comparing excellent researchers (those whose average rates of scientific publication in the last 10 years were in the top 10% of the publication ranking list) to their average counterparts, they found that the former were more active in participating in industrialization. In sum, study results showed that always there was no balanced relationship between academic research and UICs at the level of individual researchers.

To analyze the patents of UICs in Japan and Europe, we must trace information on inventors rather than simply tracking joint applications. Enterprises never independently applied for patents generated by UICs until the introduction of the American contract mode. In these cases of UICs, therefore, university researchers were typically participants in collaborative research. Lissoni (2008) found that an investigation focusing only on the data from joint applications might result in an undervaluation of the significance of innovation in universities, according to a list in a patent database of academic researchers from Italy, France, Sweden, and Denmark. Currently, there is no similar research concerning Japan.

1.2.4 Limitations of UIC in Japan

Most studies on UIC practice by Chinese scholars have focused on vocational colleges. Only at the end of the twentieth century did studies on UIC practice in universities such as Tongji University, East China University of Science and Technology, Peking University, and Tsinghua University emerge. Guangpei Zhang published *University-Industry Collaboration in Japan* in 1986, a study on UIC in Japan when Chinese scholarship began researching UIC. Later, Zongming Liao introduced the UIC modes in Japan (mutual employment systems for faculty, practice sites provided by enterprises, overseas study systems, and advanced studies of staff members sent by enterprises to universities); his study dealt mainly with education in UIC. In 1994, Zongming Liao began to publish studies on Japanese UIC in the *Tsinghua Journal of Education*. In the twenty-first century, Chinese scholars paid more attention to UIC in Japan. They began to introduce cases from specific universities in addition to studies of policy and history. Meanwhile, the sources of data changed from secondary literature to primary literature, and the analyzed data gradually became more specific and detailed.

Although Chinese scholars followed UIC in Japan just as Japanese scholars did, they placed more emphasis on the presentation of policy history and literature studies, resulting in the convergence of most literature content during data compilation and interpretation. In 2009, Zhang Chenxi used Yamagata University as an example to introduce in detail the specific practice of UIGC in Japanese universities of science and engineering in *Analysis and Enlightenment on University-Industry-Government Collaboration in Japanese Universities of Science and Engineering*.

Japanese scholars were more inclined toward micro-data analysis and empirical study; examples include Motohashi's investigation and quantitative analysis of enterprises on different scales, and the micro-analysis of knowledge production within organizations with mathematical methods in studies by Baba and other scholars.

In conclusion, studies on UIC in Japan by domestic and foreign scholars have mainly addressed quantitative analysis of micro-data and investigations of large samples. Currently, there is a lack of in-depth qualitative studies on the practice of UIC in Japan. Scholars in Japan and other countries have different classification standards for UIC modes, resulting in overly complicated results. This paper attempts to give an in-depth analysis of the forms of UIC in Japan, using multiple case studies to fill the gap in qualitative study.

1.3 The Evolution of the UIC Concept in Japan

1.3.1 University-Industry Collaboration

The concept of university-industry collaboration (UIC) first emerged in the 1950s in the United States. It is also called university-industry-research institute collaboration (UIRC) in China, and university-industry-government collaboration (UIGC) (産学官連携) in Japan when the function of government is emphasized. No standard, universal definition of UIC has been established in academic circles.

Peter and Fusfeld (1982) argued that through collaboration with universities, enterprises are able to make use of the instruments, equipment, and other resources of the universities; able to establish technical standards; and able to obtain professional instruction and the latest innovative knowledge from university professors. Geisler and Rubenstein (1989) suggested that by participating in UIC, enterprises are not only able to reduce the inherent risks in research and development, save time and costs, and develop new products and technologies, but also able to assist in solving problems in enterprises, raise productivity, and stimulate internal research and development creativity.

Bloedon and Stokes (1994) defined UIC as a research activity and plan for collaboration between higher education institutions and the industrial circle. The fees for these collaborative schemes are borne by the industrial partners. Ruth (1996) argued that universities and enterprises are two sides of the same coin, and that students can face the challenges of the twenty-first century using a combination of resources from universities and enterprises.

Kang (1997) stated that with UIC directed by organizational goals, universities, and social organizations (which are similar in nature to universities) makes joint efforts based on principles of reciprocity to ensure effective utilization of resources in order to achieve the overall goals of the collaboration. Xiao (1997) defined UIC as a collaboration scheme for professional research, product development, and research and development, for which the participating organizations pay the teachers or their students.

Jiang (2000) argued that UIC is the integration of the teaching resources in universities and the technical resources in industry, and that the collaboration strategy enables the coordination of the theory of teaching with practical applications. Lian and Ma (2001) stated that UIC usually means technical collaboration among enterprises, universities, and scientific research institutions, with the enterprise functioning as the demand side of the collaborative relationship and the university functioning as the supply side. This is a process through which universities and enterprises capitalize on their own advantages in order to jointly realize the innovation process of a technology (Xiaochuan Guo, 2001). Huang et al. (2002) suggested that UIC is an effective way for universities to initiate collaboration with enterprises in order to provide students with an opportunity to apply their theories in a real-world setting and to participate in technical practice, while enterprises correspondingly obtain strategic human resource employment benefits. Aoki (2002) defined UIC as a collaborative process by which two actors (universities and enterprises) in different fields increase their development potential through the synergistic effects generated from interaction.

Wu and Lin (2004) further argued that on one hand, UIC means mutual cooperation between enterprises and universities in an attempt to, put guiding and practical technical research into practice, while also encouraging enterprises to participate in application research in the academic world in order to cultivate research and development potential, train talent, and provide information.

In addition to definitions provided by the abovementioned scholars, UNESCO, which has promoted UIC, offers an additional definition. Both UNESCO and the EU have promoted UIC as a key task. In 1993, UNESCO passed the UNISPAR Plan (university-industry-science Partnership), which called for the promotion of scientific cooperation between universities and enterprises. UNESCO believed that collaboration between universities and enterprises would make prominent contributions to the process of industrialization in developing countries, while simultaneously encouraging colleges and universities to put more emphasis on the industrialization process in order to encourage the industrial circle to collaborate with universities, enterprises, and research institutes to stimulate development. In 1994, the EU established the European Training Foundation (ETF) with the primary goal of helping member countries develop potential human resources with the support of relevant policies outside the EU by making use of their systems of education, training, and labor markets. ETF then became the center for strengthening relationships among various UIC member states.

In China, the primary concept of UIC is UIRC," which originated from the "UIRC Development Project" jointly promoted by the former State Economic and Trade Commission, the State Education Commission, and the Chinese Academy of Sciences in 1992. Since then, the Chinese Ministry of Education has explained UIRC in both narrow and broad terms in the *Action Plan for Educational Vitalization Facing the 21st Century*. In a narrow sense, UIRC is the process of exchange and collaboration among enterprises, universities, and research institutions, based on the principles of complementary advantages, mutual benefit and development, and reciprocity; while in a broader sense, it is the dynamic integration of education with production labor and talent cultivation, scientific research, and combination of scientific and technological development with industrial production. Various scholars have also explained UIRC. For instance, Yan, Xu, and Wang (2008), as well as other scholars, characterized collaboration among enterprises, universities, and research institutions as UIRC. They argued that the purpose of the three-sided collaboration is to realize scientific and technological innovation and to accomplish the transfer of scientific and technological achievements into the arena of productive forces, and that the basis of UIRC was for the respective groups involved to realize complementary advantages and meet each other's needs by following the universal rule of combining science and technology with economics. Yang (2010) defined UIRC as collaboration among enterprises, universities, and research institutions carried out in various ways with innovation as its goal, and the industrialization of research achievements as its main method of realizing the integration of scientific research, production, and marketing, in order to gain the kinds of profits that could not be obtained by any member of the group independently. According to some scholars, however, the actors participating in UIRC do not cover the so-called scientific research institutions. For instance, Li, Wang, and An (2008) posited the definition of UIRC in terms of subjects and objects. They suggested that the subjects were the implementers (enterprises) and organizers who conducted intentional and planned technology transfer, and were motivated by scientific and technological talent and research achievements. Meanwhile the objects in question were the universities, which had a strong impact on the subjects. They argued that UIRC was a process of interaction and mutual promotion broadly encompassing knowledge discovery, proliferation, materialization, and so on.

Wang, Sun, and Shen (2008), and other scholars, argued that UIRC was the result of independent choices made by different interest subjects with distinct advantages under the conditions of the market economy.

Some scholars have objected to the formulation "UIRC." They believe that the government plays a very important role in the collaboration between enterprises and universities (or scientific research institutions), and propose to replace "UIRC" with "UIGC" (C. Wang & P. Wang, 2005). In fact, the concept of "UIGC" first emerged in Japan. In 1981, the Japan Federation of Employers' Associations began to implement the *Fundamental Technology Research and Development System for Industry of the Next Generation*, the core policy of which was intended to ensure the collaboration of government, industry, and universities in order to maximize all of their advantages. To realize this goal, Japan established the "Promotion Head Office," which would be responsible for reviewing, approving, and coordinating the basic plans for collaboration and then implementing them, strengthening information communication, and promoting research (Deng, 1989). The university-industry-government system, which brings together government, enterprises, universities, and research institutions in Japan, has played a very important role in the development of innovation capability and the rapid rise of Japan (Fang, 2002).

"Industry," in a broad sense, means private enterprises, non-profit organizations (NPOs), and other commercial organizations. Research and development in industry is directly related to economic activities, and therefore, it plays a very important role in UIGC. "University" refers to universities, the Inter-University Research Institute Corporation, higher professional colleges, and other academic institutions (national, public, and private), including comprehensive universities and specialist colleges. The basic missions of these organizations are education and academic research, and secondarily, in the wake of educational reforms, social service and contribution to society. These organizations still bear the responsibility for talent cultivation, creation of future knowledge, inheritance of human IP, and other important tasks. "Government" means national experimental research institutions, public experimental research institutions, and independent administrative institutions with research and development abilities, etc. The missions of these institutions are to realize the policy objectives, carry out fundamental frontier research based on the improvement of Japanese scientific technology, propose specific research objectives according to the policy requirements, and launch key research and development projects based on the strategic research. In addition, public experimental research institutions still bear the responsibility of technical development and instruction based on the actual needs, for instance, of regional industry. Meanwhile, local public bodies also bear the responsibility of the construction of research and development bases, the formulation of improvement policies, etc.

1.4 Study Design

Chinese academic circles pay little attention to the practice of UIC in Japan due to the difficulty of gathering research data on the practice, not because they consider the practice unimportant. Generally, it is very easy to obtain the text of policies related to UIC on a national level. Particularly in a country where information disclosure has become standard government behavior, almost all important policy texts can be obtained from the official government website. The texts cover not only the policy but also its review process. However, the participants in the practice of UIC are universities and enterprises. While they can disclose their information, they are not obliged to do so. Since UIC is generally related to enterprise secrets, their specific information is usually not disclosed. It is, therefore, difficult to study the practice of UIC.

1.4.1 Case Selection

Given that the practice of UIC is always closely related to specific actors, individual projects will focus on whether the actors in question are universities or enterprises. For this reason, there is no doubt that a case study is the most appropriate approach in this instance, but the question of which actors in practice/project to select as cases for the study can present problems. Fortunately, the author had the opportunity to study in Japan on a government scholarship in 2012, and went to Hiroshima University. Hiroshima University is only a national university, but is an important research university in Chugoku/Shikoku. UIC in this university has a relatively important role in Japan, and it subsequently became the author's first choice for the case study. After a discussion with her Chinese and Japanese supervisors, the author visited the practice site of Hiroshima University for UIC: the Center for Collaborative Research & Community Cooperation. Professor Yutaka Otsuka (大塚丰), her Japanese supervisor, contacted Mr. Matsui (松井老师) from the University-Industry Regional Collaboration Center on behalf of the author. In August 2012, after signing the "Agreement on Internship of Chinese Doctoral Students in the Center for Collaborative Research & Community Cooperation" and the "Confidentiality Agreement for Non-staff's Short-term Work in the Center for Collaborative Research & Community Cooperation," the author began working in the Center in order to translate the brochures for core products from Japanese to Chinese. Her hours and tasks were relatively flexible. The author was allowed to translate content chosen in discussions with Mr. Matsui at the Center 2–3 days/week. In fact, the author spent most of her time reading annual reports, meeting minutes, project schedules, and other internal documents about UIC at the Center, while also interviewing eight teachers at the Center in order to identify more specific research targets. After a participatory observation of more than a month, the author surveyed the UIC projects at Hiroshima University, and concluded the following: first, UICs in some subjects are frequent and deep, which make them appropriate as potential case studies. The graduate schools of science and engineering at this university include the Graduate School of Engineering, the Graduate School of Advanced Sciences of Matter, the Graduate School of Biomedical & Health Sciences, the Graduate School of Biosphere Science, and the Graduate School of Science. Engineering and the advanced sciences of matter are neighboring disciplines, and emerged in high frequency in UIC, so they could be considered as the key targets of interest. Second, while some UIC projects run smoothly, others fall by the wayside. Valuable lessons can also be learned from these failed UIC projects, but there is no sufficient information about them, and in order to consider them complete and viable cases for study. Third, some of the participating enterprises are large and famous, while others are small or medium-sized. This paper focuses on UIC in research universities, which possess obvious advantages when collaborating with famous enterprises with substantial technical strength. In addition, UICs at such universities are comparatively more complicated, and rely more on institutional systems, meaning that they have comprehensive study value. As a result, UIC projects with large enterprises are particularly appropriate objects of study. All of these conclusions greatly influenced the selection of projects for study at the University of Tokyo and Waseda University. The University of Tokyo is a representative national university, and plays an important role in promoting UIC. TLO at this university is honored as "unique TLO." Waseda University is the best private university in Japan, and its UIC is always ranked at the top for private universities. The author expected to observe the fundamental elements of effective UIC, and attempted to extract and build models of organic elements thereof through detailed study of the implementation of UIC projects at these three universities in order to lay out a path that other universities and enterprises could follow to achieve effective UIC.

The author selected the Vegetable Preservation Collaboration Project at the University of Tokyo, the Nissan Engine Development Collaboration

Project at Waseda University, and the Mazda Internal Combustion Engine Collaboration Project at Hiroshima University as case studies. She identified two channels for collecting information after selecting the universities and projects.

The first channel was the study of available literature related to the actors in each case: the author collected annual reports from 2010 and 2011, case manuals from 2011, TLO transfer specifications, and other text information issued by the UIC centers of these three universities, as well as obtaining some other information on the transfer of scientific research output from their websites.

The second channel was the interview with UIC participants: the author used the core participants in these three cases as the main interview subjects (Table 1.4). She classified them according to their affiliation, i.e., university, enterprise, or government, primarily selecting professors from universities, engineers from enterprises, and executives or decision-makers from government organizations. Given that not all UIC cases involved the government, the author interviewed government employees only for the case that involved government support. The interview outline (Appendix I) was semi-structured and was provided to the interviewees in advance. The interview focused primarily on the following four questions:

- What was the main task of the collaboration, and how did you achieve it?
- What was the most impressive occurrence during the collaboration (challenge or innovation)?
- Was your target reached upon completion of the collaboration, and if so, what did you gain from it?
- What was the influence of current industry policy on the collaboration?

The interview was the main source of material for the research results. The author studied at Hiroshima University for year and interviewed 45 people total from Hiroshima University, Waseda University, and the University of Tokyo, and finally selected 20 of the interviews to serve as the basis of the study. For important information, the author interviewed certain individuals repeatedly. The interviews were divided into the individual interviews and the group interview, and the total time of the audio recording was 1609.54 min (about 28 h). To better understand the interviewees, the author independently recorded each interview, and usually spent about seven or eight times the length of the interview analyzing the information provided by the interviewee. Occasionally this process took even longer due to the vagueness in the

Table 1.4 Interviewees' data

University	Pseudonym	Title	Unit	Years of Service	Interview Date	Interview Duration	Number of Words	Interview Language
Hiroshima University	Watanabe (渡边)	Professor of Medical Science	Graduate School of Biomedical & Health Sciences	19	2012-10-30	53:04	2130	English
	Noda (野田)	Professor of Engineering	Graduate School of Engineering	31	2012-11-10	54:09	2560	English
	Ueno (上野)	Professor of Chemistry	Graduate School of Science	8	2012-11-27	54:38	3906	English and Japanese
	Miura (三浦)	President	Pearl Dynasty Ltd.	35	2012-11-22	1:52:11	2045	Japanese and Chinese
	Kawakami (川上)	Engineer	Mazda Motor Corporation	27	2012-11-07	55:02	2038	English
	Nagashima (永岛)	President	Mazda Technical Institute	33	2013-01-18	1:30:30	4788	English and Japanese
	Takada (高田)	Department Director	Hiroshima Industrial Promotion Organization	26	2012-10-30	1:34:56	3742	Japanese and Chinese
	Li (小李)	Doctor of Engineering	Graduate School of Engineering	2	2012-11-13	56:58	5171	Chinese
	Chen (小陈)	Doctor of Engineering	Graduate School of Engineering	2	2012-11-13	1:04:49	7088	Chinese

University of Tokyo	Amano (天野)	Professor	Graduate School of Agricultural and Life Sciences	19	2012-12-05	1:35:29	5017	English
	Torigoe (鸟越)	Professor	Graduate School of Engineering	30	2012-12-14	1:33:58	6372	English
	Ato (阿藤)	UCR Director	UIC Headquarters of the University of Tokyo	1.5	2012-12-05	30:59	1294	English
	Morikawa (森川)	Director	UIC Center of ZENSHO Central Food Research Institute	5	2012-12-11	2:36:07	9596	English
	Muramoto (村元)	Consultant	UIC Center of ZENSHO Central Food Research Institute	1				
	Jin (小金)	Doctor of Agriculture	Graduate School of Agricultural and Life Sciences	3	2012-12-17	1:24:49	8314	Chinese

(Continued)

Table 1.4 Continued

University	Pseudonym	Title	Unit	Years of Service	Interview Date	Interview Duration	Number of Words	Interview Language
Waseda University	Watanabe (渡部)	Dean	Graduate School of Environment and Energy Engineering	32	2012-12-08	1:25:43	5084	English
	Kurihara (栗原)	Guest Professor; Engineer	Comprehensive Environmental Research Center, Comprehensive Research Institute of Nissan	5 41	2012-12-12	2:02:19	7769	English
	Oota (太田)	Assistant to the President	UIGC Promotion Department	13	2012-12-11	1:58:44	4351	English and Japanese
	Yano (矢野)	Department Director	Comprehensive Research Institute of Nissan	6	2013-01-28	2:18:54	6014	English
	Cui (小崔)_	Doctor of Engineering	Graduate School of Environment and Energy Engineering	8	2012-12-06	50:38	4929	Chinese
	Zhou (小周)	Doctor of Engineering	Graduate School of Environment and Energy Engineering	3				

interviewees' English pronunciation. While recording the interview, the author also wrote down important viewpoints and immediate ideas and impressions, some of which became important starting points and a basis for the theoretical framework of this thesis. The author also obtained some irreplaceable, unreproducible information in interviews with Japanese enterprises.

To obtain first-hand research in the world that was not skewed or idealized, the author had to agree with what the interviewee was saying rather than pressing them on certain points, and that tried to correct for their biases on events in two way: first, to determine which of the various explanations obtained was true based on her understanding and common knowledge, and second to explore the original articulation by comparing the interview records with different viewpoints on the same problem.

In addition to conducting interviews, the author also worked in the Center for Collaborative Research & Community Cooperation at Hiroshima University as a part-time researcher for 4 months. Her participation in UIC and observation of UIC during that period were also very helpful to understand the cases. Due to various circumstances, however, the author was only able to conduct informal, a 1-week participatory observation sessions at Waseda University and the University of Tokyo, and it was very hard for her to observe the specific projects. The author, therefore, gained only a basic understanding of the establishment of UIC management organizations.

1.4.2 Information Collection and Ethical Considerations

Every researcher attempts to maintain a certain level of objectivity. Anthropologists generally advise field workers to explain their opinions about their subjects of study to both themselves and their readers, without concealing their individual interpretations (Redfield, 1957).

In this study, two key individuals contributed assistance to the completion of the final paper: Mr. Michikage Matsui (松井亨景), the Distinguished Professor at the University-Industry Regional Collaboration Center of Hiroshima University, and Mr. Yukio Hosaka (保坂幸男), the Executive Director and Senior Consultant of SATAKE Company in Nishijocho, Higashihiroshima (the honorific title will be omitted hereafter). Mr. Matsui had already studied Chinese for 6 years when the author met him. Currently, he takes two or three Chinese conversation classes each week. He enjoyed discussing many topics with the author, and when they met and often discussed the Romance of the Three Kingdoms, Shih Chi, and other ancient Chinese classics with the author. Perhaps simply due to his love for Chinese culture, Mr. Matsui was enthusiastic about the author's study. He was an assistant researcher to the author, rather than simply a key contact person. The author often discussed a puzzle she had encountered in her study and her next steps with Mr. Matsui, who might then give her some unexpected information or cases, or introduce her to new interviewees. Mr. Matsui played a critical role in shaping the opinions of the author.

Mr. Hosaka was a key person introduced to the author by Mr. Matsui, but he was ignored at the beginning of the research process. Mr. Hosaka was initially categorized as unimportant because his project was only in intermittent collaboration with Hiroshima University, which did not comply with the author's case selection standard. Later, Mr. Hosaka actively contacted the University of Tokyo in order to get the permission for the author to visit successfully, and then introduced her to relevant professors and government staff he was acquainted with. According to Mr. Hosaka, he was surprised by the author's fluent English and communication ability, and the initiative she showed during her investigation, so he tried his best to help her to successfully finish her fieldwork. As a practitioner and researcher of UIC in Japan, he is also expected to obtain different viewpoints from the observations of an outsider, so he is looking forward for the completion of this book.

When it came to interviews with professors of engineering and medical science, due to the author's lack of knowledge in these fields, the study progressed very slowly. The author had to spend a lot of time adapting to a different culture system, meaning that she often made very little progress, even in an entire month. To get past this predicament, the author had to use her spare time to consult Chinese students who had studied engineering and other relevant specialties at Hiroshima University about the meanings of engineering terminology. After studying for more than two weeks, the author no longer felt bored while discussing specialized knowledge in interviews, and gradually began to look for more appropriate ways to ask questions about specialized knowledge. Most university professors and entrepreneurs included very technical information in their responses to questions and were unable to provide definite information to specific project activities, meaning that some questions led to valuable time being wasted during the interview sessions. A familiarity with some specialized knowledge greatly helped the author during these interviews. The interviewees may have been more willing to describe the collaboration projects in which they participated because the author understood what they did. Specialized knowledge laid a good foundation for analysis for the author to understand in broader terms the basis of collaboration concepts and activities.

In the observations and interviews, the author introduced herself as a doctoral candidate, having already related in an email sent before the meeting that her task was to prepare for a doctoral paper by interviewing university professors and entrepreneurs. All of them followed the progress of this paper, and this level of engagement from her subjects further motivated the author's enthusiasm and sense of academic mission. For convenience in analysis, the author prepared a communication log related to interview arrangements, which ultimately included 89 pages of emails, with 25,808 characters. The author overcame the uncomfortableness and shyness during this investigation and the writing of this thesis, simply due to the sense of achievement derived from the substantial work and attention of interviewees, and she gradually entered in searching for the truth and even had some fun in doing so.

To enable readers to treat this information more objectively, the author has given pseudonyms to interviewees in order to avoid being able to identify the real people who contributed to this study. The author has kept Mr. Matsui and Mr. Hosaka's real names, with their permission. The real names of collaboration companies mentioned in this paper were kept only after the permission was given by the relevant personnel. The author would like to express her special thanks to all the professors and entrepreneurs who accepted her interview requests and provided sincere help and support.

The time and place also indirectly ensured the effectiveness of the anonymous method. As time passed and the researcher departed, people naturally forgot what they had said, and information that had once been important was gradually replaced by new important information. In addition, the fact that the author wrote this book in another language enabled interviewees to provide some relatively important information more securely, thereby eliminating the lasting reservations.

2

Macro-background of UIC in Universities of Japan at the Turn of the Century

Some believe that UIC in Japan can be dated to the Meiji era because a professor from the University of Tokyo established Toshiba during that time (Branscomb, Florida, & Kodama, 2003). Objectively, this cannot be definitely described as UIC, though Japan had the good sense to introduce advanced technologies from America and Europe at that time. In 1956, the Japan Federation of Employers' Associations[1] issued the consulting report *Comments on Technical Education Reform to Meet Requirements of New Era*, and in September of the same year, the Industry Rationalization Conference of MITI issued the consulting report on the *Educational System on UIC*. In 1960, the Japan Association of Corporate Executives published its policy document on UIC. All these documents enhanced and promoted the collaboration between universities and enterprises (Geng & Liu, 2007). With fast development of the economy in Japan, collaboration between universities and enterprises really began on a large scale at the end of the 1960s, with universities and enterprises taking the opportunity afforded by MITI's university project system and other national projects promoted by the Science and Technology

[1]The Japan Federation of Economic Organizations (Nippon Keidanren in Japanese) was established in April 1948 based on the Japan Federation of Employers' Associations, which was organized in 1946 in Tokyo. It is a national membership economic organization with 47 regional associations and 58 industrial associations. The regional associations and the industrial associations are the two motivations to promote constant advancement of Nippon Keidanren, while they are independent in activities and funds. Nippon Keidanren's leadership realized that, as an economic organization, it is "a leader, not just a boss," which is also its guiding ideology. As one of the four economic organizations in Japan (Japan Business Federation, Japan Chamber of Commerce and Industry, Japan Association of Corporate Executives and Japan Federation of Economic Organizations), Nippon Keidanren clearly defines its function as an organization to solve all problems centering on people and to work for people. It believes that people are the assets of enterprises and devotes itself to three missions: (i) build-up moral consciousness of employers and establish enterprise culture; (ii) build harmonious labor relationships; and (iii) make contributions to the country and society.

Department for the progress of science and technology. In its fifth consulting report in 1971, the Japanese Science and Technology Conference endorsed strengthening organic communication between researchers from universities, national organizations, and private research institutions and making this a long-term policy for scientific and technological development. At the end of the 1970s and the beginning of the 1980s, with the strategy of "developing the country via technology" being promoted, UIC attracted more attention from the government and developed fast. In 1983, to address the challenges arising from collaborative research under UIC arrangement, the Japanese Ministry of Education established a "system for collaborative research by national universities and private enterprises" and assigned the "Research Coordination Office" at the "Science International Bureau" to promote collaboration between the industrial circle and universities.

At the three-case universities in this study, UIC began at the end of the last century. It is, therefore, necessary to understand the macro-background of UIC in Japanese universities at the turn of century. In this chapter, the macro-background will be analyzed from the perspective of national strategy, science and technology policies, reform of national universities, and changes in research and development in enterprises.

2.1 Strategic Orientation of Developing the Country via Science and Technology Innovations

After the bursting of the economic bubble in 1991, Japan found itself in an economic downturn. The International Institute for Management Development (IMD) competitiveness indicator showed the results to be: from 1989 to 1992, Japan was ranked first in international competitiveness, but fell to second in 1993 and dropped to third and then fourth in 1994 and 1995, respectively. The Japanese government realized that it would be difficult for Japan to have a leading position in the new economic growth by relying exclusively on manufacturing, and that the country must possess a high- and new-technology industry based on research innovation if it expected to have an important place in the world. Consequently, conducting strategic basic research was seen as the key to economic recovery and the maintenance of a long-term sustainable development. In 1995, Japanese National Diet drafted the *Science and Technology Basic Law* and coined the slogan, "developing the country via science and technology innovation," which was seen as an important step for Japan in establishing its National Innovation System. This law specified in

Article 5 that the government would assist public universities, public research institutions, government-operated enterprises, corporate representatives, and organizations in enriching talents, equipment, and technologies in order to promote the development of science and technology. To promote products created through UIC research in applied science and technology, the government would monitor or assist the abovementioned research implementation units in translating research outputs into production or application. The science and technology innovation and application would be the essential channel for "developing the country via science and technology." The Japanese National Diet began to compose the *Science and Technology Basic Plan* in 1996 based on the *Science and Technology Basic Law,* in which "university-industry-government collaboration (UIGC)" was significantly strengthened, thereby enhancing UIC in universities in Japan.

The first *Science and Technology Basic Plan* (the First Plan), formulated in 1996, specified to optimize procedures for approving researchers in national organizations to engage in part-time jobs to encourage them to instruct in industrial organizations in UIC. The First Plan, formulated by the Japanese National Diet in 1996, has, since 2000, allowed university teachers to do part-time work in enterprises (MEXT, 1996). The plan stipulated that in the "construction of the new research and development system", where the creativity of the researcher was considered the base and the research activities as the axis, communication among research institutions like universities and research institutes, the cooperation and exchange among central and local public organizations and private enterprises in domestic and other countries would be strengthened, and research and development would be strictly evaluated in order to realize its flexible application. The First Plan also suggested granting priority of patent implementation to private organizations that conducted collaborative research with national organizations or national contract research. To implement the First Plan, the Japanese Ministry of Education published a report titled *Construction of New System for UIC* in 1997, requiring universities to pay attention to the demands of the industrial circle, expand research collaboration with enterprises, improve university and regional facilities (for instance, build collaborative research centers), and put research achievements in practice effectively, along with giving other specific suggestions to public and private universities. In 1998, a report titled *Construction of a New System for UIC* underlined the targets of developing research achievements into patents, enhancing the flow and effective application of patent results, building and improving technology licensing organizations (TLOs), and planning for the construction of TLOs.

In 1999, the consulting meeting of the Council for Science and Technology Policy clearly stated that "social contributions" based on UIGC would be the third target of academic research pursuits in universities, and further specified and evolutionary interpretation of the intuitional reform and role of UIGC and institutional reform. In 2003, after several reorganizations, the Japanese Information Centre of Science and Technology, established in 1957, was changed into an independent administrative institution, the Japanese Science and Technology Agency (JST; Wang, 2007). In 2001, the JST of the Japanese Ministry of Education launched the "coordinator" system, and highlighted the fact that this support system for universities engaged in UIC was widely accepted by both universities and enterprises. According to the results of a questionnaire distributed by mail by the Japanese UIC Support Committee, 93% of the universities with "coordinators" believed that the system was necessary, while 5% believed that it required improvement. The surrounding enterprises and local autonomous organizations also praised this system. About 88% of them believed that it was necessary; 5% believed that it required improvement; and 3% considered it irrelevant (Wu & Zheng, 2006). A series of policies and measures based on the *Science and Technology Basic Law* all adhered to the objective of promoting scientific and technical innovation to realize the basic target of "developing the country via science and technology" through collaborative research between enterprises and universities.

After a clear-cut definition of the national strategy for the new century was established by law, the reform of administrative organizations was imperative in order to carry out new laws and policies more powerfully. This reform, carried out by the Japanese government in the twenty-first century, both strengthened and promoted UIC in Japanese universities (Xu, 2008). The highest science and technology review organization is the Council for Science and Technology Policy, which put forward the long-term planning statements for the scientific and technological development, along with the important scientific and technological strategies and plans in response to queries from the Prime Minister. At the beginning of 2001, Japan raised the status of the science and technology policymaking body, the Science and Technology Conference, to that of a council directly under the Cabinet Office, changed its name to the Council for Science and Technology Policy, and positioned it as the head office for carrying out science and technology countermeasures. The Council for Science and Technology Policy (hereafter referred to as the Council) was an important executive body that served to strengthen the policies made by the Prime Minister and the Cabinet, and directly influenced the launch of a number of important policies. As the highest leading organization in the development of science and technology, the Council was comprised of fifteen members: the Prime Minister was the Chairperson, while the other fourteen members were relevant experts and scholars from national research institutions, universities, enterprises, etc. The Affair Bureau, consisting of 70 members, was established under the Council. All these members were authoritative scholars, principals of high-tech enterprises, and government officials in science and technology. The main tasks of the Council were to conduct surveys and reviews in consultation with the Prime Minister in order to realize basic policies of science and technology development; formulate national science and technology development strategies; approve and evaluate research subjects, science and technology budgets, distribution policies for talent, and other resources; coordinate trans-provincial department affairs; etc. That reconstruction of the Science and Technology Conference significantly strengthened the science and technology administrative system in Japan and promoted the integration and collaboration of science and technology with research. The Ministry of Education, Culture, Sports, Science and Technology (MEXT) serves as the comprehensive administrative department for science and technology, and was responsible for the overall coordination of the relationships amongst the science and technology administrative departments, and also for the high-efficiency implementation and promotion of science and technology policy. Its specific work scope encompassed both efforts to reform the science and technology system and coordinating technical exchange and collaboration, not only between universities and research institutions, but also between provinces.

The First Plan was only a prelude to the reform, and did not specify clear implementation targets, but did guarantee and promote the smooth implementation of the third *Science and Technology Basic Plan* (the Third Plan) in laws and policies, administrative organizations, and personnel systems. Below, the author will discuss the national science and technology development strategy in Japan by focusing on the Third Plan.

In 2001, Japan launched the Second Plan, stipulating that Japan will win thirty Nobel prizes in the next 50 years (MEXT). Japan reaffirmed this target and established the "Research Center" in the Karolinska Institute of Sweden after its scientist, Ryoji Noyori, won the Nobel Prize in Chemistry in October 2001. To enhance the development of advanced science and technology, the Japanese government dedicated USD 130.5 billion in scientific research funds in 2000, making it second to the US in public funding for scientific research (Figure 2.1). Due to the economic depression, the total budget for the year 2002 in Japan was significantly smaller than that

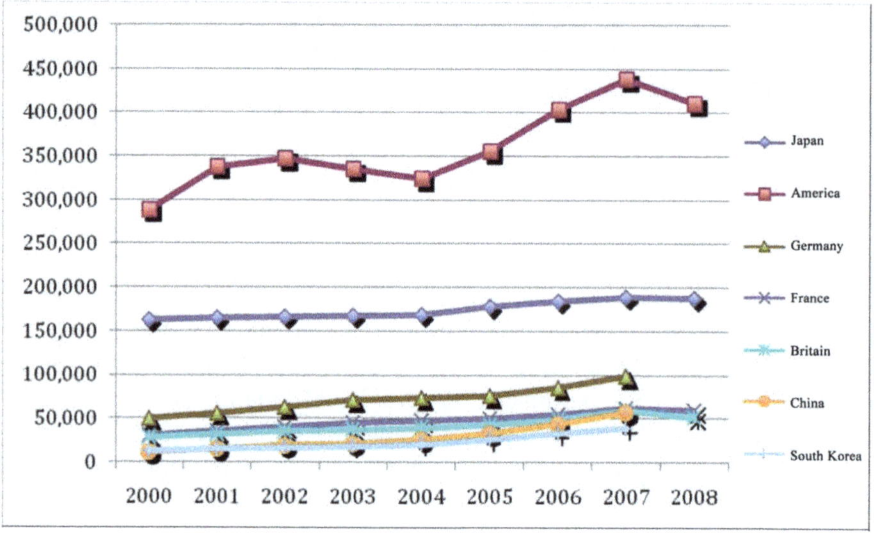

Figure 2.1 Comparison of research funds in different countries (2000–2008; unit: billion JPY).

Source: Japanese Science and Technology Agency. "Databook on University-Industry-Government Collaboration (2010–2011)", p. 57.

of the previous year, but the budget for science and technology nevertheless increased. From 2006 to 2010, Japan allocated JPY 25,000 billion (about USD 250 billion) for the government to invest in research and development. The basic concept of the Second Plan was to make Japan "a country that can create knowledge and make contributions to the world by the flexible application of knowledge," "a sustainable development country with international competitiveness," and "a country that can provide safe, secure, and comfortable life for its people" (Cabinet Office, 2004). Based on this concept, the Second Plan proposed four key development fields within the science and technology: life sciences, information communication technology, environment, and nanotechnology/nanomaterials.

In 2005, the Council, which was chaired by the Prime Minister, proposed implementing a third 5-year *Science and Technology Basic Plan,* to begin in 2006. The basic ideas of the Third Plan included conducting an "exhaustive selection and concentration" of the science and technology research subjects, promoting the "prioritization of science and technology strategies," repaying citizens with scientific research achievements, putting more emphasis on talent cultivation than equipment improvement during the distribution of budget funds, creating an environment suitable for young people to distinguish themselves and for women and foreign researchers to fully demonstrate their abilities, and making efforts to increase the proportion of women researchers to 25% during those 5 years. The Third Plan introduced another four fields of focus, adding energy, manufacturing technology, social infrastructure, and frontier science to the four key fields developed in the Second Plan (Table 2.1). This demonstrated the Third Plan's shift in emphasis to frontier science and social infrastructure that served people's livelihood.

The Fourth Plan, which was launched in 2011, was formulated with a development view for the next 10 years. It focused on "green" (environment/energy) and "life" (health) sciences, and proposed to increase the potential productive force by emphasizing technology innovation. In 2020, the proportion of government research and development investment in GDP will be increased to 1% from 0.67% in 2008, and the government–private organization collaborative research and development investment in GDP will be increased to 4% (Council for Science, Technology, and Innovation, 2011). The Fourth Plan highlights "recovery and reconstruction after the earthquake," and has human well-being as its development focus.

A comprehensive analysis of the four existing Science and Technology Basic Plans showed that the *basic research* was not changed and the strategic targets became clearer and more definite over time. The Third Plan emphasized the *talent strategy*, while the Fourth Plan updated it to a *scientific and technological talent strategy*. Meanwhile, the Fourth Plan based the creation of an *international research environment* on the *international cooperation* of the Third Plan, and emphasized the construction of an ecological environment in order to integrate *research*, "*talent*," *environment*, and *innovation*. The Fourth Plan particularly emphasized the importance of combining the efforts of universities and enterprises to realize the development of targets for the next 10 years. Before this plan, each specific target stressed the importance of collaboration by stating "implemented and promoted under collaboration and joint efforts of universities, public research institutions and the industrial circle" (Council for Science and Technology Policy, 2011, p. 9, 11, 13, 16).

In conclusion, the Science and Technology Basic Plan was the beginning of UIC in Japanese universities. It was the first fundamental law passed on science and technology, and provided legal guarantees for UIC. The Fourth Plan, passed later, specified the key collaboration fields, identities of researchers, ownership of the patent rights from research achievements, among other guidelines, and provided an operable scheme for universities and enterprises that further promoted the UIC process in universities

Table 2.1 Comparison of the four science and technology basic plans

Items \ Year	1996–2000	2001–2005	2006–2010	2011–2020
Budget amount	JPY 17000 Billion	JPY 24000 Billion	JPY 25000 Billion	JPY 25000 Billion
Actual amount	JPY 17600 Billion	JPY 21100 Billion	JPY 21700 Billion	—
Key fields		Life sciences; Information communication; Environment; Nanotechnology/ nanomaterials	Life sciences; Information communication; Environment; Ultra-fine technology (nanotechnology) energy; Manufacturing technology; Social infrastructure; Advanced science	Recovery/reconstruction after earthquake; Environment/energy; Medical treatment/ nursing/health
Strategic guidelines	Basic research; Applied research	Basic research; Applied research; Advanced research	Talent strategy; Basic research; Technology innovation; Backbone technology; International cooperation	Basic research; Talent strategy; International research environment; Science and technology innovation

Note: This table was sorted by the Author according to the information in the four *Science and Technology Basic Plans* on the MEXT website.

in Japan on a practical level. The reform of Japanese government organizations in the new century also provided an institutional and systemic guarantee of effective implementation of UIC. In addition, accompanying policies and measures help to launch a motivated collaboration between universities and enterprises. But it was the third Japanese higher education reform (National University Corporatization) that really pushed UIC in universities in Japan to a new level. The reform will be described in detail below.

2.2 Specific Impact of National Technology Licensing Policy

In 1996, the first *Science and Technology Basic Plan* proposed to grant priority in patent implementation to private organizations that conducted collaborative research with national organizations or national contract research institutes. Under the guidelines of the First Plan, Japanese Ministry of Economy, Trade and Industry formulated and issued the *Act on Special Measures Concerning Industrial Revitalization* in 1999, which was regarded as the Bayh-Dole Act of Japan, and formally came into force in April 2000 (Cabinet Office, 2010). The act shifted patent ownership from the individual to the organization level, and granted universities ownership of patents from sponsored projects. Before the issuance of this act, patents could only be owned by inventors. Since the establishment of the "Bayh-Dole system" in Japan, the contract research accepted by provincial departments has increased every year. The application rate of the "Bayh-Dole system" in all Japanese provincial departments reached 57% in 2001, rising to 94% in 2003 and to 99.9% in 2005 (Ministry of Economy, Trade, and Industry, 2007). Currently, the Bayh-Dole Act applies to all research contracts for collaboration among Japanese government entities, universities, and enterprises. In fact, research and development projects entrusted to enterprises by the government were also well received.

To further promote UIC, vigorously strengthen the basic research and development of advanced technologies, and maintain and increase the country's scientific and technological competitiveness in the world, the Japanese Ministry of Education formulated a new measure for the "distribution of patent income in national universities" in the beginning of March 2000. According to this measure, half of the income from the patents obtained by national universities (i.e., national patents) would be delivered to the state treasury, while the remaining half would be delivered to the universities, and be recorded in the universities' research budget income. (i.e., more the patents received by a national university, the greater would be its research funds.) Additionally, the Japanese Ministry of Education in 2000 abolished the regulation that universities could sign only a 1-year contract with enterprises, and encouraged universities to sign a multi-year contract with enterprises, in order to create a good working environment and favorable conditions for UIC and to promote long-term collaboration between universities and enterprises. The change in patent ownership regulations was seen as an important measure for making universities full partners in collaboration with enterprises. Since the change was made, research universities have begun to participate actively in the transfer of their scientific research achievements, and have received income from their own patents. In 2011, the patent assets of Keio University were ranked first among Japanese universities, while those of the University of Tokyo and Hiroshima University were ranked fifth and eighth, respectively (Table 2.2).

Over the course of the 1990s, the Japanese government gradually realized that the old university intellectual property management and technology licensing system was flawed, and as a result in 1998 issued the *Act on Promotion of Technology Transfer from Universities to Private Business* (TLO Act), which was the first law in Japan concerning the transfer of research achievements from universities to the industrial circle. According to Article 1, the law is aimed at promoting the transfer of technical research achievements to private enterprises, organizations, and units (individuals) from universities, higher technical colleges and the Inter-University Research Institute Corporation, and national research organizations. It does so by taking necessary measures to develop new production fields; increase technical levels within industry; and stimulate energy in research in universities, higher technical colleges, the Inter-University Research Institute Corporation, and national research organizations, in order to effect the smooth adjustment of Japan's industrial structure, the healthy development of its national economy, and the advancement of its academy (Ministry of Economy, Trade and Industry, 2011). Japan established a TLO as stipulated by the 1998 law. A TLO is an intermediary organization that is mainly responsible for the acquisition and protection of patents from the research achievements of universities and for figuring out how to industrialize these achievements with as little delay as possible. To strengthen the links between TLOs and universities, the Japanese government provided funds, equipment, and other support to nationally recognized TLOs, and also allowed TLOs to use the research equipment in national universities for free. The law included provisions

Table 2.2 Ranking of universities by patent assets (2011)

Ranking	University	Assets (Yens)	Number of Patents	Ranking	University	Assets	Number
1	Keio University	15,266	233	6	Tohoku University	11,452	309
2	Nagoya University	13,298	225	7	Kyushu University	10,966	136
3	Okayama University	12,917	111	8	Hiroshima University	9,971	194
4	Tokyo Institute of Technology	12,400	400	9	Osaka University	9,971	199
5	University of Tokyo	12,316	289	10	Hokkaido University	8,116	141

Source: (株)パテントリザルト社調査(2011年3月).

for the "transfer direction of university technology," "promotion of business for technology transfer of industrial reservation funds," "special provisions for small and medium-sized business investment and consultation companies act," and "special consideration for academic application research." The law had the effect of increasing the enthusiasm of relevant actors within UIC for obtaining patents. The *Act on Special Measures Concerning Industrial Revitalization,* implemented in 2000, additionally proposed setting up TLOs in universities, required construction of new systems for accelerating UIC in the advanced fields of science and technology, and encouraged enterprises to entrust research and development to national and public universities in the long run.

According to the definition given in a report called "University Network for Innovation and Technology Transfer," the TLO is the intermediary organization that grants patents for the research achievements. Once university researchers have obtained a patent they may agree to license it for a fee to enterprises for an agreed amount of time. Through cooperation with both academic and industrial circles, TLO puts the profits earned from the transfer of scientific and technological achievements back into scientific and technological research, and also plays a role in the "creative cycle of knowledge" (Rissanen & Viitanen, 2013, http://www.researchgate.net/publication/267995240_Report_on_Japanese_Technology_Licensing_Offices_and_RD_Intellectual_Property_Right_Issues). The Japanese government divides TLO into two classifications. The first is "recognized TLO" (Table 2.3), which is recognized by both MEXT and the Ministry of International Trade and Industry (MITI). Its business scope covers the exchange of intellectual property owned by universities and research institutes of various kinds or owned by individuals. It enjoys an annual fiscal subsidy of up to JPY 30 million, a loan guarantee with the upper limit of JPY 1 billion, and other preferential measures. "Recognized TLO" was the main form of TLO. Up to September 2010, Japan had established 46 recognized TLOs in total (Japan Science and Technology Agency (JST), 2011). Under the preferential conditions established by laws, the national universities of Japan established TLOs one after another. There were two kinds of organizational structures. One was independent of universities, as was the case with TOUDAI (University of Tokyo) TLO, and the other was an internal organization of the university, such as the TLOs at Waseda University, Hokkaido University, etc. (ORGANIZATION, 2009).

"Internal integration TLO" means the TLO is established within a university as an internal organization. It is able to make use of the capital and facilities of the university. The primary technological achievements it accepts are

Table 2.3 Forms of recognized TLO (June 2010)

		External TLO	
Year	Internal Integration TLO	External Integration TLO	Regional TLO
1998	Japan UIGC Intellectual Property Center for Universities*	(Joint-stock company) TLO of University of Tokyo	Kansai TLO (Joint-stock company) (Joint-stock company) Tohoku Technology Door
1999	UIGC Promotion Center of Waseda University*	(Limited company) Yamaguchi TLO	
2000	UIGC Exchange Center of Tokyo Denki University*	(Joint-stock company) UIC Organization in Kyushu	(Financial group) Nagoya Industrial Science and Technology Research Institute
2001	Intellectual Property Center of Meiji University*	(Financial group) Rewarding Committee for Production Technology Research TLO of Tokyo University of Agriculture and Technology (Joint-stock company)	(Financial group) Osaka Industrial Promotion Organization (Joint-stock company) Niigata TLO
2002	Intellectual Property Promotion Center of Nippon Medical School Group*	(Joint-stock company) Campus Create（キャンパスクリエイト）	(Financial group) Kitakyushu Industrial Academy Promotion Agency (Limited company) TLO of Kanazawa University
2003	Science and Technology Exchange Center of Tokyo University of Science*		(Joint-stock company) Shinshu TLO (Financial group) Hiroshima Industrial Promotion Organization

Universities marked *with asterisks* are private universities and those *without asterisks* are national universities.

Source: Homepage of Ministry of Economy, Trade and Industry. University-Industry-Government-Collaboration Policy: Technology Transfer in Universities (TLO) (2012). http://www.meti.go.jp/policy/innovation_corp/top-page.htm

the inventions of the teachers at the universities, which are the intellectual property of the universities. This kind of TLO has no corporate capacity. Its operation capital includes capital provided by "independent administrative institutions—infrastructure organizations of small and medium enterprises" in addition to that provided by the university. External TLO is established outside the university. External integration TLO means the TLO operates independently but still serves the university. Regional TLO serves a whole region. It makes use of the capital and facilities of the regional government, and provides service for enterprises and universities in the corresponding region. Established in the form of a joint-stock company, TLO is the external organization affiliated with a university. The university invests no money. This kind of TLO is invested in by university teachers or jointly with social enterprises and organizations. It accepts the technological achievements that belong to university teachers. Most of the TLOs in financial groups are affiliated with existing financial groups. They cooperate with universities to carry out the transfer of technological achievements after the reconstruction of the existing financial groups. The operation capital of TLO consists of four parts: capital provided by organizers (professors), membership fees, funds provided by "independent administrative institutions – infrastructure organizations of small and medium-sized enterprises," and income from the transfer of patents and other technological achievements.

The second classification is "approved TLO" (Table 2.4). This kind of TLO can be established just after the approval by MEXT or the responsible minister of the corresponding province. Approved TLO is generally derived from science research institutes that are supported by the government, and is mainly responsible for the transfer of national intellectual property. It can enjoy exemption from application fees, handling fees for patent registration, and registration fees, but it cannot enjoy fiscal subsidies or loan guarantee.

With the development of TLO, large transaction platforms have gathered large amounts of supply and demand data, and have been found to contain more advantages in promoting intellectual property transactions. The Japanese Patent Office has offered many patent circulation exhibitions in order to build a communication platform for technology transfer in universities and enterprises. In 2004, Japan established the University Network for Innovation and Technology Transfer (UNITT), which had 69 formal members at its inception, primarily TLOs (22 TLOs in total) and the intellectual property organizations of universities (University Network for Innovation and Technology Transfer, 2014).

Table 2.4 Establishment of approved TLO and affiliated organizations (June 2010)

Approval Month	TLO Name	Affiliated Organization	Provincial Department Managed
April 2001	(Financial group) Innovation Center of National Institute of Advanced Industrial Science and Technology of Japan Technology Transfer Association	(Independent administrative institution) National Institute of Advanced Industrial Science and Technology	Ministry of Economy, Trade, and Industry
May 2003	(Financial group) Human Science Financial group	Research organizations under administration of the Ministry of Health, Labor, and Welfare	Ministry of Health, Labor, and Welfare
June 2003	(Juridical association) Food and Agriculture Research and Development Association	Research organizations under administration of the Ministry of Agriculture, Forestry, and Fisheries of Japan	Ministry of Agriculture, Forestry, and Fisheries of Japan
April 2004	(Financial group) Assistance Center for Advanced Telecommunication Technology Research	(Independent administrative institution) National Institute of Information and Communications Technology	Ministry of Internal Affairs and Communications

Source: 特許庁ホームページ 承認·認定TLO（技術移転機関）一覧 (2012).
http://www.jpo.go.jp/kanren/tlo.htm

In 2004, barriers between national, public, and private universities were broken down after the corporatization reform of national universities; since then, all Japanese universities have introduced a competitive system of key research funding and an external assessment system for the performance. The technology transfer ability and the development of start-ups established by the universities became important factors in assessing the performance of those universities. A university cannot obtain competitive research funds without passing the external assessment, and some researchers may be eliminated from consideration for funding. To gain more competitive capital, universities make use of their advantages to proactively participate in technology entrepreneurship and strengthen education, training, and knowledge for pioneering enterprises. Venture businesses are another form of technology transfer in universities. Based on the access to advanced technologies made by professors at universities, these start-ups are established by university professors or students.

According to the report on the "Basic Investigation on Venture Businesses of Heisei 23 (2011)," up to the end of 2008, there was a total of 1,809 venture businesses established by universities in Japan, among which venture businesses established by public research organizations accounted for the highest proportion (38.7%), with the venture businesses established by professors accounting for the second highest proportion (22.5%; Ministry of Economy, Trade, and Industry, 2011, p. 1). In terms of industrial fields, research achievements in biotechnological fields were the most transferable, with 40.5% transferred into civil products, while those in IT fields were the second most transferable, with 19.5% transferred (Ministry of Economy, Trade, and Industry, 2011, p. 3).

Among the national universities, the University of Tokyo established 125 venture businesses (Table 2.5) and was ranked first, followed by the University of Tsukuba and Osaka University, while Waseda University established the most venture businesses among private universities (Table 2.6). While twelve enterprises have issued shares and been listed above at present, approximately ten other enterprises are slated to be listed this year. In the future, the number of listed enterprises may reach 180.

According to the statistical data on the establishment of TLOs and pioneering enterprises in universities, the TLO Act not only lays a legal foundation for higher education in Japan to carry out UIC and provides legal guarantees and support to UIC in higher education, but also greatly expands the development scale of UIC on a practical level.

Table 2.5 Number of venture businesses established by national universities (2008)

Ranking	University Name	Number	Newly Established	Ranking	University Name	Number	Newly Established
1	University of Tokyo	125	4	6	Tokyo Institute of Technology	57	1
2	University of Tsukuba	76	4	7	Kyushu University	55	5
3	Osaka University	75	0	8	Kyushu Institute of Technology	45	1
4	Kyoto University	64	0	9	Hokkaido University	43	2
5	Tohoku University	57	1	10	Hiroshima University	38	1

Source: Ministry of Economy, Trade and Industry. Practice Report on "Investigation of Foundation for Entrepreneurial Ventures in Universities," (2009), p. 15. http://www.meti.go.jp/policy/innovation_corp/whatsnew/fy20vn.pdf#search='平 成+20+年度大学発ベンチャーに関する基礎調査

Table 2.6 Number of venture businesses established by private universities (2008)

Ranking	University Name	Number	Newly Established	Ranking	University Name	Number	Newly Established
1	Waseda University	74	6	6	Tokai University	20	1
2	Keio University	51	3	7	Digital Hollywood University (デジ・タルハリウット・大学院)	19	3
3	Ritsumeikan University	35	1	8	Kochi University of Technology	17	0
4	Ryukoku University	27	0	9	Doshisha University	16	0
5	Nihon University	23	1	10	Tokyo University of Science	15	1

Source: Ministry of Economy, Trade and Industry. Practice Report on "Investigation of Foundation for Entrepreneurial Ventures in Universities," (2009), p. 17.

2.3 Reform in Universities Centered on National University Corporatization

In the process of UIC between Japanese universities and enterprises, national university corporatization is the reform movement motivating the promotion of UIC from the point of view of the university. The reform mechanism of national university corporatization was first proposed at the beginning of the 1970s. In October 2003, the National University Corporation Law was formally enacted, following disputes and discussion that lasted about 30 years. Given the duration of the reform effort, it is clear that this reform has played a very important role in the development of higher education in Japan. For the purpose of Article 17 of the *Science and Technology Basic Law,* formulated in 1995, the government introduced science and technology personnel as required and formulated laws to appropriately extend the limitation on employment of national public servants in order to improve the process of the science and technology personnel introduction system, and prepare open and fair qualification review methods. This article clearly defined the direction of national university corporatization.

2.3.1 Reform Centered on National University Corporatization

Before this reform, national universities were always established as subsidiary bodies of government and were managed by the central government. The US Education Envoy Screw once characterized the national universities of Japan as "protected bureaucracies" (Osaki, 1997, pp. 151–163). In managing internal organizations, if a university wished to establish or discontinue an organization, it had to apply to MEXT and establish or dismantle the organization under its supervision. With regard to financial matters, MEXT determined the amount of university funding to allocate and how the funds would be used. All these measures limited the ability of universities to undertake free education and research: the government was excessively protecting universities while also strictly limiting them. Half of the funding for these universities was government allocations, making national universities too dependent on decisions made by MEXT, causing them to lose power and independence, and rendering them unable to meet the needs of societal development in a flexible manner.

As national public servants, teachers had no incentive to research projects that could be beneficial to the society because neither their pay nor their jobs depended on it. University teachers and management personnel paid no attention to repaying the taxpayers. A former president of a national university believed that teachers at national universities had almost no thought of repaying taxpayers, and most were engaged in education or research based on their own interests, meaning that the education continued to be dominated by the pursuit of knowledge for its own sake. Moreover, the University of Tokyo and the six other national universities were specially prioritized by the government in the national budget, meaning that their motivation to create income was low. The Asahi-Shimbun (newspaper) reported that local national universities and private universities could not be motivated, either (as cited in Tian, 2009, p. 30).

At the end of the 1990s, people in the industrial circle began to criticize the isolation and ossification of national universities. They believed that these national universities worked at a low rate of efficiency and could not provide the talent and service required for the development of the society and economy. Some scholars indicated that, although "learning" was considered an important component of economic success in Japan, almost no achievements made by the nation after the Second World War could be attributed to the higher education (Hayes, 1997). The industrial circle also publicly acknowledged that "they preferred to import technologies from American universities to relying on Japanese scholars" (Miyoshi, 2000, pp. 669–697). The Japanese Ad Hoc Council on Education strongly condemned higher education organizations on the basis that they lacked individuality, that their research was at too low a level to be internationally recognized, and that universities were essentially closed and ossified organizations that were unable to adapt to the necessities of social and international competitiveness.

With increasingly intense competition in the global economy since the beginning of the twenty-first century, the industrial circle in Japan has begun to turn from the manufacturing of the 1990s to the fields of information and advanced technology. In order to beat out competitors, enterprises were in urgent need of support and collaboration from universities for the development of innovative solutions. The economic bubble of the 1990s made enterprises unable to afford the "lifetime employment system" in self-cultivation, and expected to hire graduates from universities to provide services for them. In addition, enterprises turned to self-dependent innovation of technology instead of the introduction of technologies and patents from the US and Europe. Therefore, "the question of how to escape the current situation of weak capacity in basic and original research in Japan became critical" (Hu, 2004, pp. 97–102). In the face of a crush of innovation in advanced technologies, enterprises expected to realize technical transformation and innovation via collaboration with universities.

After the national university corporatization reform, collaboration between universities and society became closer. To increase the sustainable development potential and the international competitiveness of Japanese enterprises, one measure of national university corporatization was to make collaboration between national university corporations and enterprises closer, and encourage national university corporations to provide more technologies, patents, and talents as required by enterprises. This reform was also aimed at gaining more sources of capital for universities, thus reducing the financial burden on the government.

After the national university corporatization reform, the Japanese government enacted measures to promote collaboration between universities and enterprises, as well as other social organizations. Collaboration between universities and enterprises made the relationship between universities and society closer, which promoted the social development by taking advantage of universities. It was not only required by the development of the current society and economy, but also a core mission essential to universities. The Investigation Committee for the National University Corporation emphasized in discussion, "each university shall do business under UIGC (collaboration of industrial circle, academic circle and government departments) in a flexible and effective manner" (2002). According to Article 22 of *The National University Corporation Law*, national university corporations shall accept commissions from non-corporations and perform education or research activities with them; furthermore, universities shall give open lectures to provide learning opportunities for outside personnel. The National University Corporation Law substantially promoted the development of UIC in universities at an institutional level.

Since the national university corporatization reform, one of the key indicators in assessing a university was the institution's contribution to society. With this incentive, universities and professors were inclined to engage in applied research and short-term projects in areas that could directly produce social benefits, such as nanotechnology, life sciences, and the automobile industry. They were less willing to participate in basic research and time-consuming projects that they liked, but probably, that would not directly produce economic benefits, such as philosophy, literature, and pure natural science. In other words, this evaluation system violated academic freedom.

With regard to finances, national universities received special attention from the government, which guaranteed stable funding for these universities to some extent. Even if the income of a university decreased, its teaching, research, and other activities would not be significantly impacted, because government allocations always accounted for the majority of the annual budget of the university. Private universities, on the other hand, did not enjoy such treatment. Their funding depended to a large extent on their own efforts, and was obtained from a variety of sources. Therefore, after the national university corporatization reform, both private and national universities faced the same severe conditions in competing for research funds, and thus became more active in UIC.

2.3.2 Changes in Internal University Organizations

The organizational structure of universities also determines the degree of UIC. Under the strategy of "developing the country via science and technology," the establishment of a new organizational structure within universities and the reform of the internal structure of the Japanese government played important roles in enhancing the intensity of UIC.

To guarantee that the research collaboration system between national universities and the industrial circle succeeded, the Japanese Ministry of Education followed the example of the Industry–University Cooperative Research Center Program of the National Science Foundation in the United States, set up the Cooperative Research Center Program in 1987, and began to establish "Cooperative Research Centers" in national universities as places to further promote scientific and technological collaboration between universities and the industrial circle (Nie, 1997). With national university corporatization, the relationship between the Japanese government and the national universities changed from a direct subordinate relationship to the more contractual one promoted by UIC in universities (Xu, 2008). Some studies, by comparing the number of collaboration projects between universities and enterprises before and after national university corporatization in 2004, have found that the national university corporatization mechanism further stimulated collaboration between universities and enterprises (Lu, 2006).

Some people believed that the legal definition of university teachers by the government limited UIC. The Japanese National Diet passed part of the bill of amendments to the *Law for Special Regulations Concerning Educational Public Service Personnel* in 1996, guaranteeing that the pensions of the teaching and administrative staff who temporarily left their positions to participate in UIRC would not be affected after they retired, and breaking the legal obstruction to interaction between university personnel and the industrial circle. In 1997, the Japanese government instated legal documents that would permit university teachers of a civil servant nature to do part-time work in enterprises (Tang, 2007). Such government-formulated policies related to research funds were the key to guaranteeing university teachers would participate in UIC. The Act on Special Measures Concerning Industrial Revitalization exempted universities and teachers from patent application fees, encouraged university teachers to take part-time jobs in enterprises, provided financial aid, and took other measures to guarantee teachers would participate in UIC in financial support (Xu & Li, 2006). To create a free environment for entrepreneurs and reduce limitations on teachers' entrepreneurial activity due to institutional factors, Japan modified the *Law for Special Regulations Concerning Educational Public Service Personnel* in 2000 to conditionally ease restrictions on part-time work for researchers from national universities and scientific research institutes. Since 2004, due to the process of national university corporatization, Japan has eliminated the national civil servant status of teachers and staff at national universities, making it possible for researchers from national academic organizations to participate in science and technology innovation in industry.

In June 2001, MEXT put forward "reform policies for (national) university structure"[2] (also known as the Toyama Plan). The main focus of which was the acceleration of the industrialization of university research achievements, with specific targets; for instance, the number of patents converted into enterprise research outcomes increasing to 700 from 70 within 5 years, and the construction of more than 10 Silicon Valley-type high-tech industrial parks within 10 years (as cited in Hu, 2004, p. XX).

In addition to the obvious effects of national policies on the promotion of UIC, UIC policies formulated by universities directly promoted the comprehensive development and ranking of universities. For instance, the Tokyo University of Agriculture and Technology, which was not a key university, made substantial achievements in UIGC by formulating UIC policies based on its actual demands. Its ranking in comprehensive indicators increased from twelfth in 2006 to third in 2007 (Xu, 2008).

Strict UIC management organizations established in universities are the institutional guarantee that universities will participate in UIC. Some Chinese scholars believed organizations in universities were improved to promote UIC after studying UIC organizations at the University of Tokyo. But given that the program at the University of Tokyo had only been in place for a relatively

[2]Because this policy was proposed by Toyama Atsuko, who was then the Minister of State for Science and Technology, it was called the "Toyama Plan" as well.

short amount of time, its long-term effects were not yet clear (Wang, 2007). Other scholars believed that well-established organizational structures such as the "UIC Head Office," established by the University of Tokyo in 2004 (Xu, 2008), could promote UIC in universities.

2.4 Research and Development Forms in Enterprises and Institutional Change

The forms of research and development in Japanese enterprises have undergone significant change since the end of the last century.

2.4.1 Technology-Introduction Stage, Led by Large Enterprises (1950–1989)

As early as the 1950s, technological advancement in Japanese enterprises was attributed first and foremost to the introduction of overseas technology. In 1950, the Japanese government issued the *Foreign Investment Law* and the *Foreign Exchange Act* as part of their strategy of introducing overseas technology. According to statistics, Japan introduced, on average, 103 new technologies per year from 1950 to 1959, 469 new technologies from 1960 to 1967, and, after the autonomization of trade, capital, and technology, up to 1,061 and 1,154 new technologies in 1968 and 1969, respectively (Zhang 1992). With this spike in the introduction of new technologies, Japanese competitiveness in science and technology accordingly grew. At the end of the 1960s, excluding the military and space fields, Japan had introduced almost all significant projects that applied advanced technologies produced since the 1930s in the United States and Europe (Zhi, 2008, p. 70). Therefore, the industrial technology level of Japan in the 1960s had generally caught up to that of the United States and Europe.

Clearly defining the ownership of intellectual property in a system created by the government would further encourage enterprises to pursue knowledge innovation. Xu and Li (2006) indicated that the Japanese government passed the *Act on Special Measures Concerning Industrial Revitalization* in 1999, enabling a portion of nationally-owned intellectual property to be returned to enterprises (instead of being solely owned by government) in order to encourage enterprises to be more active in UIC.

During this stage, comprehensive research funds came mainly from private enterprises and government. Table 2.7 shows that private enterprise investment accounted for about 70% of research funding, government investment 25–33%, and foreign investment only 0.1%.

Table 2.7 Japanese research and development funds in proportion from 1965 to 1978 (Unit: JPY 100 million/%)

Year	Total R&D Funds	Government Investment		Private Investment		Foreign Investment	
		Amount	Proportion	Amount	Proportion	Amount	Proportion
1965	4258	1312	30.8	2944	69.1	3.1	0.1
1970	11953	3014	25.2	8935	74.7	4.3	0.1
1975	26218	7206	27.5	18993	72.4	17.8	0.1
1980	35700	9995	28.0	25674	71.9	30.6	0.1

Source: Prepared based on relevant data from the *White Paper on Science and Technology* issued by the Japanese Science and Technology Agency, 1980.

The users of research funds were mainly private enterprises, various research organizations, and universities. In terms of usage proportions, Table 2.8 shows that in 1965, private enterprises used 59.3% of the total funds, while universities and research organizations used 40.7%. In 1970, private enterprises used 68.9%, while universities and research organizations together used only 31.1%. From 1972 to 1976, due to the impact of the global oil crisis, the proportion of research funds used by enterprises decreased slightly to 63.4%, but it rose again after 1976.

Private enterprises in Japan are strongly independent due to their leading position in investment in research and development and in use of research funds. In addition, Japanese enterprises, especially large enterprises, had implemented an isolationist "research and development at their own expense" strategy for a long time, so that they were not active in UIC from the Second World War to the 1980s.

At the end of the 1970s, the technical gap in important fields among Japan, the US, and Europe grew narrower, which meant that the era of wholesale introduction of technologies from the US and Europe was over. Meanwhile, the captious market in Japan had various and complicated requirements of industrial technologies. Around that time, the Japanese government had come to the realization that the scientific system established after the Second World War, based on "developing the country via trade," worked against the development of innovative science and technology, and that a research and development system led by private enterprise hindered original research and failed to cultivate creative talent. By failing to cooperate closely with universities and national research institutes, private enterprise was to a large extent impeding the generation and cultivation of the "seed" of innovative science and technology (Wang, 1998).

With the deepening of economic globalization in the 1980s, the competition for products in international markets became more intense. Large enterprises' long-term strategy of research and development at their own expense showed lower efficiency. Establishing a closer research and development collaboration relationship with universities was an obvious way to compensate for the insufficiency of research and development resources in enterprises and, at the same time, increase the efficiency and quality of research and development. As a result, enterprises strengthened collaboration with universities and public scientific research institutions starting in the 1980s. According to the *Field Investigation Report into External Collaboration of Enterprises under the Innovation System in Japan in 2003* by RIETI, 71.4% of Japanese enterprises had some kind of research and development

Table 2.8 Use proportion of research funds by different sectors from 1965 to 1978 (Unit: JPY 100 million/%)

Year	Total R&D Funds	Private Enterprises		Various Research Organizations		Univeristies	
		Amount Used	Proportion	Amount Used	Proportion	Amount Used	Proportion
1965	4258	2524	59.3	684	16.0	1050	24.7
1970	11953	8233	68.9	1546	12.9	2174	18.2
1975	26218	16848	64.3	4207	16.0	5163	19.7
1976	29414	18822	63.4	4715	16.6	5877	20.0
1977	32335	21095	65.2	4943	15.3	6297	19.5
1978	35670	22910	64.2	5634	15.8	7126	20.0

Source: Prepared based on relevant data from the *White Paper on Science and Technology* issued by the Japanese Science and Technology Agency, 1980.

collaboration with one or more external scientific research institutions. The main UIC modes were collaborative research, contract research, scholarship contribution, and a mutual dispatching of researchers. The original form was like a constant debtor–creditor relationship between certain enterprises and universities, while UIC was more like supplementing the research institutes with enterprises. These modes had played active roles in the past, but presented clear barriers to obtaining required knowledge for new enterprises and SME. Therefore, the new enterprises and SME were eliminated from these constant collaborative relationships (Aoki & Harayama, 2005).

2.4.2 Large Enterprises and Small- and Medium-Sized Enterprises Playing Equal Roles in UIC

With the bursting of the economic bubble in Japan and the development of economic globalization in the 1990s, the lifetime employment system, which supported the constant development of large enterprises for many years during the period of high-speed development of the economy, began to break down, and small and medium-sized enterprises became "active."

For a long time, large enterprises had played a dominant role in the establishment of the National Innovation System in Japan, while the status of small and medium-sized enterprises was low, and their effect weak. For research and development, large enterprises had always made use of internal research institutes and other research resources of their own, and consequently were passive in UIGC and other external collaborations. The method of "research and development at their own expense" in large enterprises is enterprises' R&D strategy, also known as NIH Syndrome, was the largest obstruction to UIGC (Motohashi, 2003, p. 2). The main reason for these conditions was that the long-term research and development strategy in Japanese enterprises did not match the UIC promotion strategy. But after the issuance and implementation of the Second Plan in 2001, large enterprises began to actively explore and perform various research and development collaborations at deep levels. According to the *Investigation on Current Situation of Collaborative Research with External Organizations* by RIETI, both the types and scales of enterprises engaged in external collaboration increased greatly compared with rates of external collaboration 5 years before. Especially given the constant development of economic globalization, networks, and informatization since the 1990s, an innovation system that relied on "research and development at their own expense" was no longer able to adapt to contemporary conditions. In the context that information technology, biochemical and medical technologies, and nanotechnology had gradually come to lead innovations in science and technology around the world, Japanese MEXT launched the strategy of Developing the Country via Science and Technological Innovation. With the great aim of a "new Japan to lead the world modes," Japan established several strategies at the national level to be implemented successively, including Developing the Country via Intellectual Property, Developing the Country via Talents, Developing the Country via Innovation, etc. In this way, UIGC was provided with a strategic support from science and technology, talent, intellectual property, and innovation. According to Articles 18 and 19 of the *Science and Technology Basic Law,* the government must take preferential measures in land tax and other levies and some preferential policies in financial support in order to promote research and development in private enterprises, and must also provide necessary support for private research and development that complies with the targets of the national science and technology development plan.

The operations of small- and medium-sized enterprises began to evolve, with diversified products, multi-angle collaboration, and independence. These enterprises mainly aimed at special markets and put more emphasis on maintaining originality in technology such that they were able to develop a variety of innovative products. Following this trend, small and medium-sized enterprises increasingly attracted more attention. The number of people who were engaged in research and development and research funds in these enterprises increased at a stable rate. Meanwhile, with the increase of funding for research and development, the number of research achievements rapidly increased accordingly. According to the *Investigation on the Actual Situation of Innovation Activities Conducted by Enterprises* by the Small- and Medium-sized Enterprise Agency, in the 5 years from 1996 to 2000, the number of patents applied for as a result of the research activities of large enterprises increased to 10,651 from 8,648, an increase of 23%, while the number of patents applied for in the manufacturing industry by small and medium-sized enterprises increased to 2,489 from 1,754, an increase of 42% (Figure 2.2). This clearly demonstrates that the status of small and medium-sized enterprises in UIC had increased. The proportion of collaborations by universities and other scientific research organizations with large enterprises decreased from 87.3 to 66.6%, a rate of more than 20%, while that of small and medium-sized enterprises increased from 12.6 to 33.4% over the same period (National Institute of Science and Technology Policy, 2011).

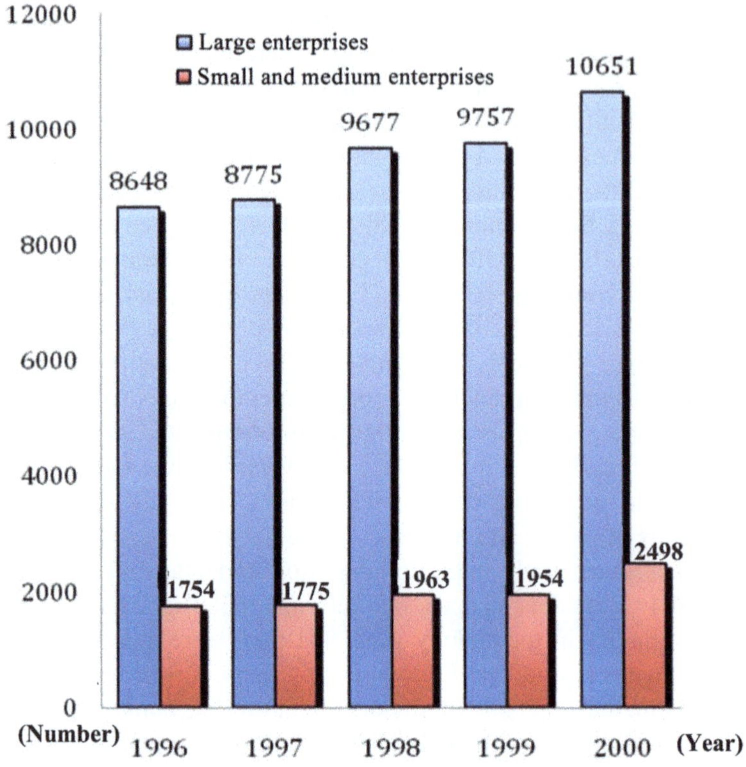

Figure 2.2 Patents applied for by manufacturing enterprises of different scales from 1996 to 2000.

Note: "Small- and medium-sized enterprises" means enterprises with less than 300 staff members.
Source: The Small and Medium-sized Enterprise Agency. *Investigation on Physical Forms of Reform in Operation of Small and Medium-sized Enterprises* (2001).

Generally, enterprises increasingly based the collaboration with universities on three main objectives: exploration of new development fields, development of cutting-edge technology, and development of basic research that embodied the advantages of the knowledge innovation system represented by universities. Under this framework, Japanese enterprises became more and more active in external research and development collaboration. In 2003, MEXT conducted an investigation of research activities undertaken by 868 private enterprises and compared the proportion of external research and development activities in the periods from 1998 to 2003 and from 2003 to 2008, as shown in Figure 2.3 and Figure 2.4, respectively. According to the

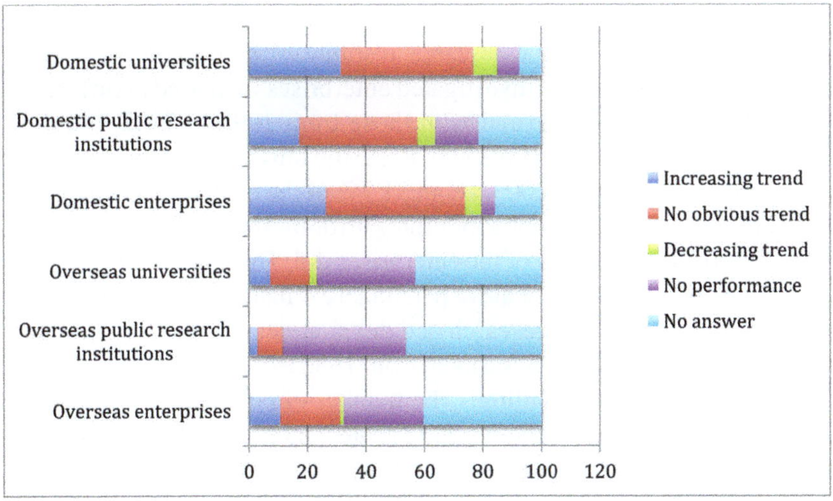

Figure 2.3 Investigation on intention of external collaboration in Japanese enterprises from 1998 to 2003 ($N = 868$).

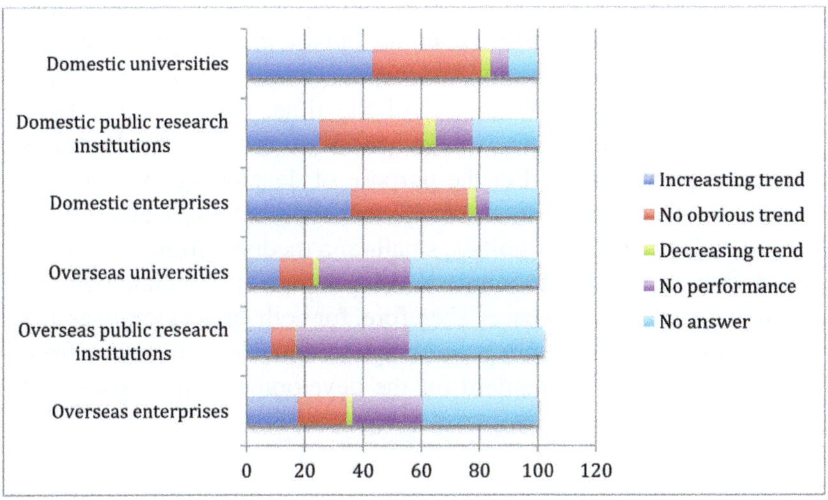

Figure 2.4 Investigation on intention of external collaboration in Japanese enterprises from 2003 to 2008 ($N = 868$).

Source: www.mext.go.jp, Innovation under Knowledge-based Economy.
(http://www.mext.go.jp/b_menu/shingi/gijyutu/gijyutu11/houkoku/05042301/021/004.pdf# search=
'知の時代を先導するイノベーションの創出'

two figures, the intention to carry out external collaboration with domestic and overseas universities, public research institutions and enterprises, and other collaboration actors in the investigated enterprises increased at different levels from 2003 to 2008 compared with those from 1998 to 2003, but the intention of enterprises to collaborate with domestic universities increased the most drastically. From 1998 to 2003, about 31.7% of the enterprises increased their research and development collaboration with domestic universities, while from 2003 to 2008, about 43.3% of the enterprises did. The number of UICs in Japan increased quickly, and the scale of personnel exchange expanded greatly. According to the statistics, the research and development funds from private enterprises accepted by universities tripled from 1986 to 2003, increasing from JPY 25.98 billion to JPY 83.4 billion. The number of collaborative research projects between national universities and private enterprises increased from 56 in 1983 to 8,022 in 2003, and the number of contract research projects increased from 1,286 in 1983 to 6,987 in 2003 (Zhi, 2008).

Large enterprises once dominated the industrial circle in Japan, but by the beginning of the twenty-first century, collaborative research with small enterprises was booming. The "research and development at their own expense" approach of large enterprises highlighted their strong scientific research capabilities, but after the bursting of the economic bubble in the 1990s, and with the approaching knowledge explosion, independent research and development could no longer meet the requirements of the time. Consequently, large enterprises established collaborative research modes with research universities for resource complementation. For the purpose of developing its economy, Japan began to focus and support small- and medium-sized enterprises. For seeking development opportunities, small- and medium-sized enterprises collaborated with universities to jointly develop unique new technologies to guarantee their stable survival space. Therefore, for both large enterprises and small- and medium-sized enterprises, seeking strong technical support from universities had become the standard for the development in the industrial circle.

3

Collaboration between University of Tokyo and ZENSHO Group

Food Preservation Project, Collaboration between Graduate School of Agricultural and Life Sciences and ZENSHO Group

3.1 Introduction

The University of Tokyo is Japan's leading university, and plays a very important role in the country's higher education landscape as a whole. It is also the university that first carried out UIC in Japan. The literature shows that the Toshiba Corporation was a venture enterprise established by professors at the University of Tokyo in the early stages of the development of UIC (Branscomb, Florida, & Kodama, 2003).

The author repeatedly sought possible channels for investigating UIC at the University of Tokyo after arriving to Japan in July 2012, but failed to gain access in her first 4 months of residence. In November, she was lucky enough to be accompanied by Mr. Matsui of the Center for Collaborative Research & Community Cooperation at Hiroshima University while visiting the SATAKE Group. Yukio Hosaka, the Executive Director and Senior Consultant of SATAKE, graduated from the University of Tokyo and obtained his doctorate degree in agriculture there. When the author said she planned to select a case of UIC from the University of Tokyo, Mr. Hosaka offered to contact professors at the university with UIC experience.

This was a welcome surprise, and as a result of Mr. Hosaka's introduction, the author was able to go to the University of Tokyo. When he learned that the cases the author planned to study concerned engineering, life sciences, and science, Mr. Hosaka contacted Mr. Amano. After several rounds of communication, Mr. Hosaka scheduled a visit on a day when he was attending a meeting in Tokyo so that he could attend his meeting in the morning and introduce the author to Mr. Amano in the afternoon. At noon on December 5, Mr. Hosaka and the author met at the gate of the Graduate School of Agricultural and

Life Sciences at the University of Tokyo, and went to Mr. Amano's research room. On the way, the author told Mr. Hosaka that she hoped to visit the UIC Center of the University of Tokyo. After meeting Mr. Amano, Mr. Hosaka asked if he would contact some people at the UIC Center for a meeting. Unexpectedly, Mr. Amano contacted them by phone at that time. After talking with the author for less than 20 min, Mr. Amano set a time for the author to meet with Mr. Ato, the Director of the Division of University Corporate Relations (DUCR). After the interview with Mr. Amano, the author met Mr. Ato, accompanied by Mr. Hosaka. Unfortunately, due to Mr. Hosaka's tight schedule, this meeting ended abruptly. Nevertheless, Mr. Ato had prepared a lot of internal information related to UIC at the University of Tokyo for the author, which significantly supplemented the interview. Later, the author stayed in Tokyo for 1 week and visited the UIC Center at the University of Tokyo multiple times.

People involved in this chapter are briefly introduced below:

Professor Amano: He studied at the University of Tokyo for his bachelor's, master's, and doctoral degrees in agricultural mechanics. After graduating with a doctorate in 1985, he taught in the School of Agricultural Sciences at Mie University. In 1994, he returned to the Graduate School of Agricultural and Life Sciences at the University of Tokyo. Professor Amano had visited France and Belgium for 1 year each in the year 1987 and 2000, respectively.

Jin: A Chinese student studying in Japan who graduated from Jilin Agricultural University and earned both her master's and doctoral degrees in Food Science at the Graduate School of Agricultural and Life Sciences at the University of Tokyo. She mainly studied near-infrared spectrophotometry. During her master's study, she gradually shifted her focus to the link between science and food.

Ms. Ato: The director of DURC, having taken this position a year and a half earlier. Before working at DURC, she worked in the Department of Architecture in the School of Engineering. The term of the DURC Director is 3–5 years. Generally, a new director is hired after about 3 years.

Mr. Morikawa: A staff member at the UIC Center at the ZENSHO Central Food Research Institute. He obtained his bachelor's degree in biology from Waseda University and then obtained his master's degree in genetics from the University of Tokyo. Mr. Morikawa joined ZENSHO in 2008 and worked successively in the Sukiya and Udon sub-branches of the Financial Management Department. In 2009 he joined the Central Food Research Institute, where he was responsible for connecting collaboration projects with universities.

Muramoto: Obtained his doctorate at the University of Tokyo. He worked at the Patent Office for 2 years in Tokyo, starting in 2006, and held the post of DURC Director from 2008 to 2011. Through promoting UIC at this university, he became familiar with ZENSHO, and ultimately took the post of Senior Consultant at the ZENSHO Central Food Research Institute in 2011. He was the "boss" of Mr. Morikawa.

3.2 The University of Tokyo and ZENSHO

The University of Tokyo is the originator of UIC in Japanese universities. It has collaborated on scientific research with numerous enterprises, national research organizations, government entities, and other university departments. To help readers clearly understand the collaboration project between the university and ZENSHO, the author will give an introduction to the development of UIC in these two organizations below.

3.2.1 Collaboration Tradition at the First National University

The University of Tokyo was the first national university established in Japan, and is one of the oldest universities in Asia. It was established in 1877, at the beginning of the Meiji Restoration, at the direction of the Japanese Ministry of Education, and was renamed as the Imperial University in 1886. This university has always played a significant role among Japanese universities and the higher-education community (Amano, 2011). In 1880s, many schools constituted the University of Tokyo's predecessor, many foreign scholars were introduced to their faculties, and tried to set up organizations specialized in producing the special skills so urgently needed by both enterprises and the government (Bartholomew, 1989). The predecessor to the University of Tokyo's Department of Engineering was the Engineering Department, established in 1873. This department has always been active in participating in collaboration with the industrial circle, most notably through faculty member Professor Fujioka's (藤岡) establishment of an enterprise (the so-called start-up venture) that developed into the company now known as Toshiba (Odagiri, 2003).

The University of Tokyo began their technology transfer and other similar activities very early. Internal and external organizations promoted UIC and technology transfer related to research outputs at the university. In 2001, before national university corporatization in Japan, faculty and staff at the university began to discuss UIC methods, and in 2002 created the UIC Promotion

Planning Office. In 2004, the university finally reorganized the UIC Head Office (Figure 3.1). This reconstructed UIC Head Office has been used ever since.

The UIC Head Office consists of the Division of University Corporate Relations (DUCR), Division of Intellectual Property, and Division of Industralization. As TOUDAI Technology Licensing Organization, TOUDAI TLO is a part of the Division of Intellectual Property. The original name of TLO was the Center for Advanced Science and Technology Incubation, Ltd., abbreviated to CASTI. The University of Tokyo Edge Capital Co., Ltd. (UTEC) is the joint-stock company with the university.

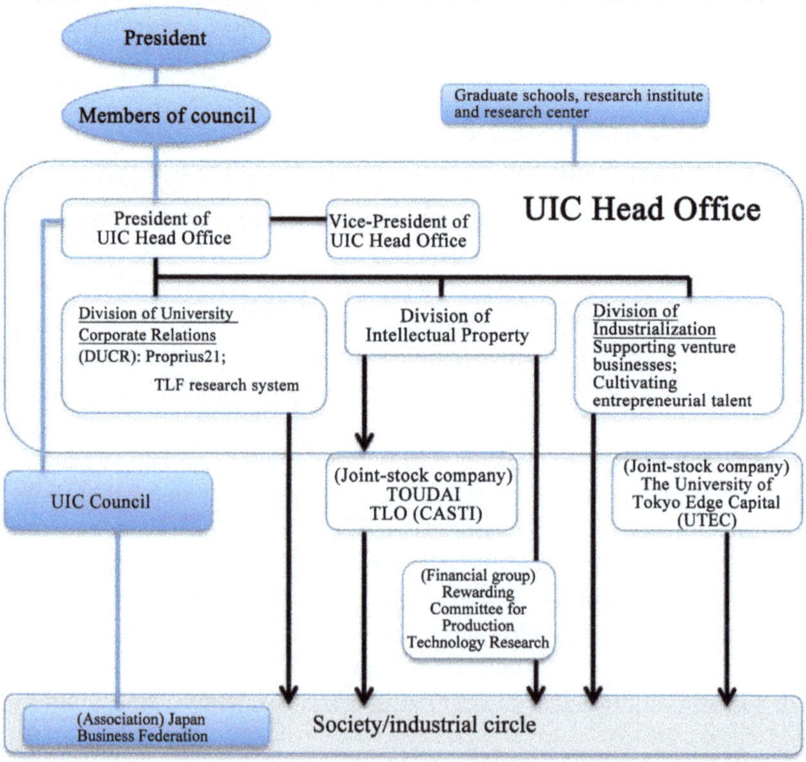

Figure 3.1 Structure of UIC at the University of Tokyo.

Source: Page 2. Division of University Corporate Relations. (2011). *Annual Report 2011*. Tokyo: University of Tokyo.

DUCR is broadly responsible for contacting universities regarding UIC inside and outside the university. To promote wider collaborative research in UIC, the Proprius21 Program (a collaborative research project setting-up scheme) was implemented in June 2004. Enterprises were able to present their intended collaborative research proposals to the DURC, including information on subject selection and collaboration scale. Then DUCR may select appropriate candidates for cooperation from the Researcher Database according to information provided by enterprises about demand, or collect collaboration proposals from all researchers at the university by publishing the report on demand of enterprises' requirements via internal information platforms (Tamai & Miyata, 2007). Before the commencement of collaborative research, enterprises discuss their research proposals, plans for collecting and dispatching results, the duration of project implementation, and other matters with researchers from the university. In January 2005, the University of Tokyo established the UIC Council (Figure 2.1), which had a total of 500 member enterprises by May 2006 (Tamai & Miyata, 2007). According to 2011 data published by the University of Tokyo, although the number of collaborative research projects continued to increase, the number of return avenue showed a decreasing trend after 2009 (Figure 3.2).

The Technology Liaison Fellow (TLF) training system is an educational training project implemented by DUCR in the beginning of April 2008. The 1-year program was designed primarily for full-time trainees dispatched by local governments. This training project was aimed at cultivating the ability of trainees in the development of UIC relationships, and enabling these

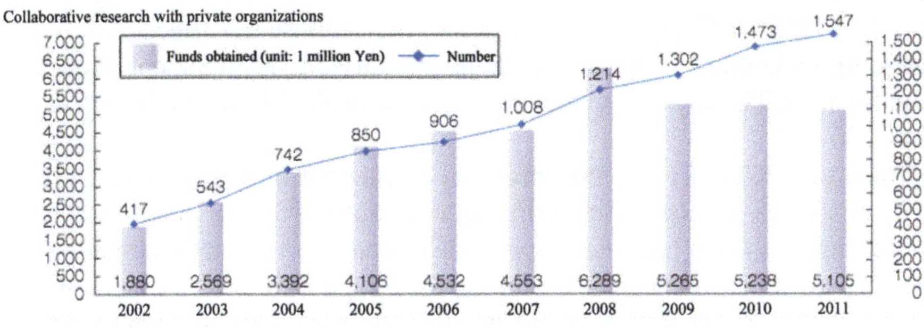

Figure 3.2 Number of collaborative research projects and funds obtained.

Source: Division of University Corporate Relations. (2011). *Annual Report 2011*. Tokyo: University of Tokyo, 2011, p. 22.

trainees to share what they had learned immediately upon their return to the workplace. This program not only nourished an active enterprise–university–government relationship by making full use of local resources, but also effectively promoted the development of industry and the local economy.

TOUDAI TLO (the former CASTI) was established in August 1998 (Division of University Corporate Relations (DUCR), 2011), one of the first TLOs to appear after the issuance of the TLO Act. It was a typical external TLO. After the corporatization of national universities in April 2004, CASTI was renamed TOUDAI TLO, and the office was moved from Marunouchi to the Hongo campus of the University of Tokyo. TOUDAI TLO set three targets for its mission of repaying society through collaboration between academics and industry, with a focus on fostering a safer society: (i) seek to collaborate with enterprises with promise and development potential to put the knowledge of the University of Tokyo into practice, (ii) make full use of the potential of collaboration to function as a bridge between the university and enterprises to ensure that the University of Tokyo maintains a close relationship with industry, and (iii) provide teachers with a good venue for negotiation to minimize contractual and legal problems and other issues in the process of patent application. Since 2004, the contract of bailment and signing of collaborative research contracts at the UIC Head Office of the University of Tokyo has been under the management of the Division of Intellectual Property (Figure 2.1). Patent applications for research achievements are the responsibility of TOUDAI TLO. Enterprises pay business commission and proprietary rights implementation fees when signing contracts. According to the Annual Report from the UIC Head Office of the University of Tokyo in 2012, the number of patents applied for by TOUDAI TLO after corporatization was almost ten times that of before corporatization (Table 2.3). Its interest/driven development mode eventually made TOUDAI TLO the most successful TLO in Japan (Tamai & Miyata, 2007).

Similar to TOUDAI TLO, the University of Tokyo Edge Capital Co., Ltd. (UTEC for short) is an organization that operates outside the university. Since its establishment, UTEC has been responsible for the effective implementation of new capital input, management and support, capital flow, etc. Currently, UTEC implements policy in three main areas related to enterprise management: first, UTEC provides growing enterprises with advantageous capital, jointly participates in the management of the first batch of risk capital through

"UTEC limited partnership 1"[1] to increase the enterprise's value, provides management services for accelerating capital flow, etc. Second, UTEC attempts to clearly define the venture's next potential investment targets through "UTEC limited partnership 2" in order to ensure a constant and smooth capital circulation chain within enterprises. UTEC can bring an expected or a more unique value to enterprises via effective management. Third, UTEC maintains a close collaboration with universities as an external organization to construct and develop an ecological system that can provide capital continuously. After the generation of the original scientific seed or technological concept, UTEC tries to identify and develop investment prospects. UIC at the University of Tokyo developed quickly due to the mutual coordination of DUCR, TOUDAI TLO, and UTEC. These three organizations made significant contributions to the construction of collaborative relationships, technology transfer, and licensing, supporting venture enterprises by making use of their own organizational advantages.

3.2.2 Transition of ZENSHO

The ZENSHO Group, established in 1983, is a famous catering enterprise in Japan. Throughout its 30-year development, it has been listed in the Tokyo Stock Exchange. The ZENSHO Group has more than thirty brands and 4,459 chain restaurants, which are mainly located in Japan, the US, China, and Brazil (ZENSHO, 2011).

The total turnover in the catering service industry amounts to JPY 24,000 billion in Japan, and the total turnover of ZENSHO is up to JPY 400 billion, accounting for less than 17% of the market share. Therefore, ZENSHO's next objective is to become the world's leading catering company, occupying 20–30% of the market share in the world (like Japan's automobile and electronics industries), and keep a rising status in international market (ZENSHO, 2011).

ZENSHO's aim is to provide safe, delicious, high-quality, cheap food. In Japan, ZENSHO possesses a unique Mass Merchandising System (MMD) with a uniform standard that covers the procurement of raw materials, processing in factories, and dispatching products to chain restaurants. This complete supply system brings huge benefits to ZENSHO. In collaboration with raw-materials suppliers, ZENSHO has put various precautions in place, testing rules and assessment standards to ensure the safety of food, and thereby

[1]"UTEC limited partnership 1" was established and implemented in 2004, while "UTEC limited partnership 2" was implemented in 2009.

reflecting the management principle of giving priority to "safety," "quality," and "cost" adhered to by ZENSHO.

In 2005, ZENSHO established a Department of Food Safety to ensure adherence to its principle, "Safety is the top priority." It was among the first enterprises to implement food safety monitoring. In the same year, ZENSHO began to provide funds to the Alliance for Global Sustainability (AGS) at the University of Tokyo.[2] According to Mr. Morikawa, ZENSHO was in the process of consulting about whether to lay foundation for a collaborative research project on food safety measures with the University of Tokyo at that time.

In 2006, ZENSHO set up the Central Food Research Institute within its Department of Food Safety. It became the first enterprise with an internal research institute in the catering service industry. The institute was responsible for analyzing the chemical composition of pesticide residue. In 2007, ZENSHO established a monitoring system to test the safety of local agricultural products and the products of animal husbandry. The test results were the basis for future raw-material import policies. Moreover, the Central Food Research Institute began to use radiation dosimeters to test the radiation in food materials dispatched to all chain restaurants. By collaborating with universities, the Central Food Research Institute was able to begin its research into formulating international food safety standards. In order to improve its technology development abilities, ZENSHO participated in DUCR's Proprius21 Program at the University of Tokyo in 2009 to seek out professors and researchers who met its technical requirements. Since then, ZENSHO has formally begun its collaborative research with universities. After starting in 2009, ZENSHO finished its first collaborative research project with the University of Tokyo in 2011, a 2-year collaborative project focused on microbiological detection development. With the support of the Proprius21 Program, ZENSHO will continue to strengthen its collaborative research with the University of Tokyo, as technical development has become an important pillar of its development.

Mr. Muramoto worked at the Central Food Research Institute as a senior consultant, and was mainly responsible for building its laboratory, designed for collaboration with the university. The new laboratory was built in April

---

[2]Alliance For Global Sustainability of University of Tokyo: consists of University of Tokyo, Massachusetts Institute of Technology, Swiss Federal Institute of Technology, and Chalmers University of Technology. The four unviersities, which represent Asia, North America, and Europe, carry out joint research to promote development of countermeasures for the environmental protection and sustainable development.

2012 and was in trial-run stages at the time of the author's visit. The two parties signed only a 1-year contract. Mr. Muramoto's goal was to support and promote future technology research for ZENSHO by ensuring the stability of the collaboration laboratory.

3.3 Vegetable Preservation Collaboration Project

The vegetable preservation collaboration project between the University of Tokyo and ZENSHO is a research project at the initial stage of collaboration. Although the University of Tokyo has extensive experience in UIC and has completed many successful collaborations, in its initial stages, any collaborative research project can come up against a running-in and trial process.

3.3.1 Launch of the Project

ZENSHO began to pay attention to the technology development relatively recently. Since its establishment in 1983, ZENSHO has regarded food dispatching, chain restaurant expansion, and other organizational management modes as its primary objectives; food testing, preservation, and other relevant technologies were not included in the development schedule of the family-type restaurant chain. But as a result of the increasing demands for taste and mouth-feel by consumers, ZENSHO created its Department of Food Safety to develop technologies related to food preservation in 2005, and began to provide scientific research funds to the Alliance for Global Sustainability (AGS) at the University of Tokyo. In 2009, ZENSHO participated in DUCR's Proprius21 Program at the University of Tokyo in order to create a document for its future technological research collaborations.

The vegetable preservation collaboration project was established between ZENSHO and Professor Amano as a result of the Proprius21 Program. This project was implemented in 2010. It was intended to be a 3-year project, but was completed only after 1 year.

As the DUCR Director at the University of Tokyo at that time, Mr. Muramoto took the position of senior consultant at the ZENSHO Central Food Research Institute, which was connected through the Proprius21 Program. The Central Food Research Institute is a research institution within the Department of Food Safety. Mr. Muramoto had a mutual understanding with ZENSHO through communication what was done for the vegetable preservation collaboration project from 2010 to 2011, and was employed by ZENSHO in 2011 to be responsible for the construction of its research and development laboratory. In 2009, Mr. Muramoto, received a memo from ZENSHO explaining that ZENSHO expected professors at the University of Tokyo who were developing technologies related to food preservation to engage in collaborative research with them. ZENSHO's application was published on DUCR's information website, and professors from across the university could apply to participate in the project according to individual interest. Finally, Mr. Muramoto sent a list of collaboration candidates (including Professor Amano) to ZENSHO. ZENSHO selected Professor Amano based on information provided about him. After establishing this relationship, Mr. Muramoto scheduled several meetings for the two parties. Professor Amano and ZENSHO spent approximately 1 year in discussing the research objective of their collaboration. ZENSHO wanted him to sign a 3-year contract, but Professor Amano would only agree to a more conservative 1-year contract.

Jin, a doctoral candidate in Professor Amano's laboratory who had enrolled in 2009, had already conducted a study related to food preservation, so she selected the vegetable preservation project as the subject of her doctoral thesis and participated in the implementation of this project. From the fall of 2010 to the summer of 2011, Jin was primarily responsible for ZENSHO's vegetable preservation project, and, together with a senior student, completed all experiments related to the project. During her collaborative research, Jin reported her experimental data to Professor Amano regularly and discussed the helpfulness of the research results. When Jin's ideas fell short, Professor Amano always required her to find answers to her problems independently via experimentation. But as time grew limited, Professor Amano contributed his own speculations and ideas, requiring Jin to verify them. Finally, Professor Amano reported the research results to ZENSHO, or ZENSHO sent staff to visit him for an update on the progress of the project. Generally, when Jin prepared the experimental data, ZENSHO sent staff to discuss it with her. This usually happened about once in every 2 months, but occasionally happened several times in a month.

During the collaborative research process, the biggest problem that Jin faced was experimental samples. Jin's research plan would be disrupted if she could not get samples in time. Generally, the research results were reported in the third week of the month. In order to keep to this schedule, Jin typically conducted experiments in the first week, analyzed the obtained data in the second week, and reported her results in the third week. But if she could not get the required sample from ZENSHO in the first week, Jin had to do the experiments in the second week, and then analyze the data overnight in that week so that she was able to report the results in the third week. As a result, in communicating with ZENSHO, in addition to formal meetings and discussions, Jin used email frequently. Before each experiment, Jin would send her research plan and schedule to ZENSHO via email to ask if and when the factory could provide the required samples. She would typically get the research samples after five or six such emails. Especially when she could not get samples in time, Jin had to do more such work.

ZENSHO operates hundreds of fast food restaurants, and the head office dispatches products strictly according to the ordered quantity submitted by each chain restaurant. The customer flow volume in the catering industry is not fixed, so the ordered quantity is different every day. As a result, Jin could get the experimental samples only when there were surplus products. The head office's vegetable factory had to wash and pack the vegetables before sending them to each chain restaurant, where they could then be directly unpacked for use. The dispatched quantity of vegetables was almost exactly the same as the ordered quantity. For this reason, Jin had to inform ZENSHO about the quantity of samples and the scheduled time for the experiment 1 month in advance in order to ensure smooth experiments. Then, before the experiments could begin, Jin generally had to spend 2–3 days confirming the samples she needed, and would usually have to wait 1 week before she received the sample arrival date via email. She often could not get a reply from ZENSHO immediately. Samples were reviewed and sealed by staff at different levels within the organization even if the technical department had approved them. Of course, given that ZENSHO operated the most fast food restaurants of any catering company in Japan, operating Yoshinoya, Sukiya, Setou Udon, Hamazushi, and other chain restaurants, Jin could understand the existence of such a complicated approval procedure for sending out samples. That said, once it became clear that she could not get samples from ZENSHO at the time of an experiment, Jin had to adjust her plan and reset the experimental time according to ZENSHO's schedule: whether the samples could be provided or not in time was a key factor affecting her research progress.

In order to better analyze their experimental data, Professor Amano and Jin visited one of ZENSHO's factories. Factory visits were among the approved data-gathering activities for the collaboration. This research platform provided universities with a chance to understand the practical reality of running an enterprise. During this factory visit, Professor Amano and Jin were admitted to some restricted-access areas that no unauthorized personnel were allowed to see, since unprepared visitors introduce bacteria that would corrupt the preservation treatment of the vegetables. Through these on site visits, Professor

Amano and Jin were able to more accurately determine the factors that affected vegetable preservation, and which furthered the progress of their research.

After 1 year of collaboration, Professor Amano published his research findings in an academic conference, with ZENSHO's permission. Finally, the two sides ended their collaboration due to a disagreement over the handling of the research results, but stipulated that they might collaborate with each other again in the future.

3.3.2 Introduction of DUCR

The Proprius21 Program was a special project of the UIC Center's promotional activities at the University of Tokyo. It was designed mainly to find researchers for collaborative research projects based on the requirements set forth by enterprises. The collaboration between Professor Amano and ZENSHO was realized with the support of this system.

At the time the Proprius21 Program was begun, Mr. Muramoto was the DUCR Director at the University of Tokyo, and he established the collaborative relationship for the research project. He sent the guidelines for the University of Tokyo's Proprius21 Program to all the qualified enterprises in Japan, including ZENSHO. He recalled that ZENSHO had been interested in collaboration, and provided them with a list of potential candidates, including Professor Amano, after searching in the university's Researcher Database. Mr. Muramoto arranged a meeting for the two parties to exchange ideas. Finally, ZENSHO decided to collaborate with Professor Amano.

According to Professor Amano's recollection, the situation before the formal collaboration began was as follows:

"We spent about a year discussing the collaboration intention and target. At first, a DUCR teacher contacted me for this, and then ZENSHO sent staff to discuss it with me. An agreement would be signed for collaboration, and the reputation of the university would be affected if the results were not ideal. Therefore, we repeatedly discussed what their research aims were and what kind of scientific research we could provide. ZENSHO expected to sign a three-year research agreement, but I was not sure about the results after research of one year's work, and rejected the proposal of signing the three-year agreement."

After a year-long discussion, Professor Amano signed the contract with ZENSHO under the auspices of the Graduate School, and the food preservation project formally began. DUCR's task ended when the project was launched.

Mr. Ato, the current DUCR Director, explained to me that the responsibility of DUCR is to establish relationships between enterprises and university professors, and it does not participate in the collaborations once these relationships are formed. Therefore, the professor and the enterprises are jointly responsible for maintaining their collaborative relationships after they have signed their contracts. Mr. Morikawa was only a participant on the ZENSHO side, not the principal of the project. He recalled that they had had four meetings in total with Professor Amano, and had also gone to his research laboratory for a face-to-face meeting in order to describe their objectives and give him their schedule for realizing these goals.

The DUCR's introduction led to the collaboration between Professor Amano and ZENSHO, but the subsequent collaboration process and their continuous collaboration was a result of their motivation in completing the project and required effort on both sides. Of course, in a good collaboration environment, enterprises and universities are better able to realize their collaboration targets (Allen & Taylor, 2005).

3.4 Win–Win and Conflicts

3.4.1 Breakthrough in Enterprise Development Bottleneck

ZENSHO, as the largest catering group in Japan, realized that technology was the key force it needed to break through the development bottleneck that hindered its progress toward becoming the biggest catering group in the world. Given that it possessed no development team, ZENSHO sought support from universities. Mr. Morikawa recalled that the company had no basic research center, and, therefore, thus could not undertake basic scientific research. The senior consultant, Mr. Muramoto, further explained:

> We had not needed technology in the past. But now, we hoped to develop ZENSHO into one of the biggest enterprises of the catering industry in the world, so we were eager for technological support. Although the Central Food Research Institute was established in 2006, we still lacked relevant research ability, so collaborating with universities turned out to be an important direction for our development. Based on this development strategy, we began to consider the technologies that could be developed; for instance, food safety testing and preservation. We plan to develop technologies as required in the catering industry, though these technologies may not necessarily be complete and perfect.

Mr. Morikawa described ZENSHO's initial development targets: "the development target of this company stays the same. Since its establishment, we have required technical support related to food, but the company was in developmental stages at that time. Now, however, we are developed, and have enough space to carry out scientific research with universities."

Mr. Muramoto underlined the transition in consciousness needed for an enterprise to have a breakthrough in development:

> Japanese enterprises generally increase their competition via strengthening internal technological research and development. But the knowledge-based economy develops so fast that we have to change our concept. We should absorb various advanced technologies from universities, other enterprises, and even overseas enterprises. This is a strategy to optimize economic investment. Only with such a strategy can an enterprise speed up its development. In addition, the operation situation of an enterprise dictates the direction selected for development. Only those enterprises with relatively high profits are able to expect to expand their development space and then turn to technology.

This is the general pattern for enterprise development. When the capital scale reaches a certain level, enterprises will consider exploring new development channels. Development of new technology naturally becomes their first choice, so they are attracted to universities with significant research achievements. Therefore, carrying out collaborative research with universities and building united laboratories is viewed favorably by enterprises.

3.4.2 Perfect Combination of Theory and Practice

In reflecting on his collaborative research with ZENSHO, Professor Amano acknowledged that he had requested and been given a lot of practical information about the production process, for instance, "the process applied for disinfection, the number of times of washing vegetables, and the water used before and after cutting the vegetables." Of course, this kind of information about the enterprise's internal workings was confidential. But Professor Amano still believed that it was a very valuable experience.

> I am able to understand the actual situation from a broader perspective. At ordinary times, we always stay in the laboratory to do experiments, and do not know what will happen when our

> knowledge is put into practice. When asking people in enterprises about actual information, we hardly get answers, for that information cannot be disclosed publicly. Mr. Hosaka is my predecessor. He will answer my questions when I ask him, and I have learned a lot of things from him. In this collaborative research, ZENSHO presented me some confidential information.

Collaborative research not only enables students to be close to practice, but also lays a foundation for their future work. The biggest benefit for Jin was the chance to become familiar with and understand the process of putting technology into practice within an enterprise. This was very valuable for a student without work experience. She was able to visit ZENSHO's factory because it was necessary for the research of ZENSHO's vegetable preservation project. This factory was not open to public. Moreover, the factory did not allow visits by individuals, because casual access would bring in a lot of bacteria, adversely affecting the preservation of fresh vegetables. But through collaboration, Jin was able to observe the processes to which her research would be applied, and, therefore, be more prepared for her future job. In her preliminary PhD viva voce examination, which she completed recently, some professors asked how her research could be implemented in real life. As she had observed the actual operating process directly, she explained the process to the professors, and how her technology was applied in practice in combination with engineering theory. After concluding this collaborative research, Jin not only finished her doctoral thesis in 3 years, but also felt prepared for her future job.

Mr. Amano commented, "Although students cannot disclose what they see in the enterprise and what they learn from discussion, they can learn something important and will understand that the subject selected for their doctoral thesis has academic significance, and, more importantly, practical significance." The organic combination of theory and practice helps students understand the practical value of dry, sometimes abstract theory, and actively stimulates students' intellectual growth in the academy.

3.4.3 Difficulty in Collaboration

Mr. Morikawa took the post of Director of the Central Food Research Institute in 2009. At that time, the institute had only been established for 3 years, so everything had to be started from scratch. He encountered significant difficulties:

> It is very difficult to start a collaboration project, because we must first clearly define the requirements of the company, then communicate with professors and adjust their targets accordingly. We should define a collaborative relationship first through a discussion with the dean and then repeatedly with professors, going over the research targets, length of time needed, estimated expenditure, and other details. Beginning a collaboration is, therefore, the most difficult part of the process. This may be because our company does not have much experience in collaboration, and additionally, we ourselves also have no research experience.

Every enterprise that initially establishes a collaboration may encounter similar difficulties. Once a long-term partnership is established, after working together for several years, each side will have established resources obligations and partnership roles, making the frequency and intensity of conflicts decrease (Tornatzky & Bauman, 1997). Continuous collaborative relationships are helpful in reducing the exchange costs to the two actors in a UIC.

But in this case, after the 1-year contract for the project was completed, the two sides did not continue their collaboration, though both Professor Amano and ZENSHO said that they still keep in touch, and may continue the collaboration in the future. Mr. Morikawa indicated that ZENSHO could not make use of the vegetable information obtained from Professor Amano's research, but it seems that in reality the two sides had divergent views on the handling of that research. Mr. Muramoto further explained:

> If the research is promising, we will consider continuing the collaboration and then apply for patents. But the current results did not show any potential, so we temporarily ended the collaboration. In any case, we need to put lots of time and capital into collaborative research, and obtaining patents is our objective. University professors still consider research and teaching as their primary objectives, so they will publish these with students to make their research public, but once the information is published, we can no longer apply for patents. For the research from this collaboration, we did not insist on confidentiality and permitted them to publish. We will consider the direction of the next collaboration only on the condition that we can obtain patents. Intellectual property is very useful for us.

As UIC is an inter-organization relationship, the two sides may have differing and even conflicting goals (Boris & Jemison, 1989). Regarding the issue of not continuing the collaboration, Professor Amano explained, "Our final target is the same, but the enterprise pursues more practical things, while I am more interested in finding and studying the basic data." In the collaborative research for the vegetable preservation project,

> We mainly test the ATP content in vegetables to check their freshness. Our test technology has been able to determine the ATP content in vegetables, but the ATP content is also affected by transportation and packaging. They believe this technology can be applied to the packing production line, but I can get more reliable data after testing and research on the packed vegetables. Of course, it is OK if they want put this technology into production now.

Professor Amano did not believe that they temporarily ended the collaboration due to conflict: "It is just because we have different views on how and when the technology should be put into practice." Although both sides were positive about the temporary suspension, the obvious conflicts in culture and concept are clear obstacles that two sides must overcome in a collaboration (Siegel, Waldman, Atwater, & Link, 2003).

Collaborative research starts from a long, complicated discussion process involving multiple actors. The participants express their intention of collaborating in some areas, and the boundaries of the areas become clearer with the deepening of the discussion. If the participants know each other and treat each other honestly, the progress of the collaboration will be accelerated (Bardach, 1998), but it is hard to meet these two conditions at the same time. If the management personnel do not have enough confidence in each other, collaborations can be supplemented by an increase in resources dedicated or a very high-quality guidance process. Similarly, if participants do not sufficiently understand each other, more time and energy can be put into overcoming the obstacle that may result.

3.4.4 Limitations on the Growth Circle of Scientific Research

Even if Professor Amano had been rewarded financially or with prestige by the collaboration, he was more inclined to a more open scientific research collaboration. But the risk of an open scientific research is very high, meaning that enterprises generally will not choose it as a main collaboration objective. Professor Amano was grateful to Mr. Hosaka for giving him this collaboration opportunity.

> Mr. Hosaka's company provided funds for my research project. At that time, Satake Corporation (Mr. Hosaka, in fact) did not clearly specify the research subject. This was a favorable arrangement for me. Satake Corp. only required me to submit a one-page rice research plan, and I was free to decide its specific contents. I could make use of the funds to carry out some important basic research on, for instance, the design of food safety standards. In fact, this kind of basic research was very important not only to this company but also to the industry as a whole.

By contrast, in his collaboration with ZENSHO, Professor Amano had to carry out research according to the specifically defined targets. Although he could accept collaboration in both modes, creative scientific research was more attractive to him. He believed that the purpose of research should be to produce basic research results for the public. He further emphasized that the research should serve the rice, not an individual or a company. From his point of view coming from the perspective of universities, knowledge should be featured with commonality, so research serves science, not individuals. The indistinct commonality is obviously in conflict with the competitiveness of the market.

Professor Amano also said, "If Mr. Hosaka's company can provide him and other researchers with funds for open research, the long-term development of his company will benefit. The company will not suffer losses due to this, and on the contrary, will benefit more from open research than from pure collaborative research on developing products. But it is a pity that many enterprises do not realize this."

In addition, on the cultivation of doctoral students, Professor Amano held a similar view:

> For unsatisfactory experimental data and results obtained by students, I usually require them to find answers by themselves. But in this collaborative research, I had no time to wait for them to find answers by themselves, so I had to give tips for possible answers. But these tips were only based on previous experience, and I did not check them with experiments. But if the students can find the answer by themselves and then check the answer with experiments, they can find the right directions. This waiting is crucial to their growth.

Time is crucial to the students for their development in scientific research and in the academy. Professor Amano expected that enterprises could provide funds for relatively long-term basic research, and he believed that the enterprises would benefit most from this kind of research. It is inevitable that enterprises will push for their own development and growth, thus seeking outside facilitation can only improve this.

3.5 Summary

In terms of institutional guarantee, UIC projects at the University of Tokyo are provided with impeccable organization, complete professional teams, sufficient policy support, etc. With the establishment of CASTI, in accordance with the Act on Promotion of Technology Transfer from Universities to Private Business, in 1998, DUCR in 2003, and UTEC in 2004, the University of Tokyo has essentially finished the construction of the base organizational frame for UIC that facilitates smooth implementation of UIC in hardware-rich facilities. The UIC Building was completed in 2004 and the Entrepreneur Building in 2007; together these two buildings provide a guarantee of necessary hardware and physical facilities. After the completion of the physical facility, the creation of a team of personnel is constructed. The internal team in the UIC organization at the University of Tokyo is large and professional. The UIC Building has eight floors. The term of office for a staff member is generally 3–5 years. Generally, the administrative personnel who have been engaged in contracting work for UIC in the university, or enterprise engineers who have rich experience in UIC, are able to apply for jobs in the UIC organization. Notwithstanding the short term of office, a high professional level is required for the UIC team to ensure the quality of UIC promotion. The University of Tokyo has also launched a unique UIC promotion policy and scheme. The university began to implement the Proprius21 Program in 2004, the "Entrepreneur DOJO," a project to cultivate student entrepreneurs, in 2005, and launched "comprehensive collaborative research projects with enterprises" through DUCR in 2006.

The ZENSHO case is very representative of the beginning stages of successful UIC cases at the University of Tokyo. However, since the collaboration did not continue, this case may be considered unsuccessful. During the case selection for this study, it was extremely difficult to find an unsuccessful case in Japan. In this humble country, very few people are willing to offer unsuccessful cases to foreign scholars for the study. The author found one such a case due in some measure to random chance. In any event, this case aligned with her initial conception of the project: seeking out the root of problems in unsuccessful cases. Mr. Ato, the current DUCR Director, observed, "The UIC cases at the University of Tokyo that we can currently count account for less than 20% of the UIC cases in the whole university." This shows that the DUCR Director is not primarily responsible for contacting universities for individual UIC projects in the university. In addition, the fame of the University of Tokyo ensures that professors will find appropriate partners without intermediary administrative agents. Of course, the influence of graduates of this university cannot be ignored. To put it simply, this case study is rare. The author believes this case may show even how a solid institutional system may have its shortcomings.

The initial stages of the collaboration between ZENSHO and Professor Amano were typical of collaborations between universities and enterprises. ZENSHO regarded UIC as a transitional strategy for future development meaning they had no previous UIC experience, while Professor Amano lacked an economic incentive for his work on the project. Given that this UIC was conducted by two inexperienced parties, it was unsurprising that the process generated a divergence. DUCR ended its introductory role at the signing of the contract, and provided no other subsequent support or guarantees. This indicates that, to some extent, the University of Tokyo is only interested in the numbers and ultimate results of collaborations, but ignores the effectiveness of the actual implementation of UIC. It can be seen that UIC at the University of Tokyo only appears to be successful, and functions primarily as an ornament decorating the university. Of course, on the other hand, the main function of DUCR was not intended to introduce one professor-to-one enterprise projects, but rather was large interdisciplinary and inter-agency UIC. It is extremely difficult to investigate such UIC projects with multiple actors. Moreover, as Mr. Ato noted, "Our work is to build a relationship, and we will not be responsible for the project once the relationship is set, so we are not able to keep in touch with relevant participants." This remark further defines the status and function of UIC in the development of the University of Tokyo.

4

Collaboration between Waseda University and Nissan Motor Co. Ltd.

Engine Project Undertaken by Graduate School of Environment and Energy Engineering

4.1 Introduction

Waseda University is a typical private university. To figure out if UICs involving national universities and private universities are comparable, the author included Waseda University in the study scope, while selecting the University of Tokyo as a case study.

After searching for about 2 months, the author finally decided to use the collaborative research between Hiroshima University and Mazda Motor Corporation as a case study, and soon contacted Zhou, a doctoral student of Professor Watanabe's at Waseda University, through a recommendation by Chen, a doctoral student at the Graduate School of Engineering at Hiroshima University. During the phone call, Zhou said that Professor Watanabe expected the author to speak to him directly for the interview. On December 6, 2012, the author contacted Professor Watanabe for the first time, then got his reply confirming the schedule for the visit, and went to Waseda University to visit him the next day. Everything went so smoothly that it exceeded the author's hopes.

Professor Watanabe was a good conversationalist, and the interview went very smoothly. The author discussed her research plan with the professor and expressed her hope to interview an enterprise in collaboration with him. Professor Watanabe recommended Nissan Motor Co. Ltd. (hereafter referred to as Nissan), given that collaboration with Nissan was relatively open, while most of the other companies kept their collaborations confidential. He then introduced the author to Professor Kurihara, an engineer who had worked at Nissan for 40 years and was currently a guest professor at Waseda University. Professor Kurihara's dual identity had many advantages. From his point of view, his experience as both engineer and professor made him particularly insightful. Finally, Professor Kurihara helped in contacting some researchers at the Nissan Research Center for the author to interview and to finish her investigation of the project.

People involved in this chapter are briefly introduced below:

1 Professor Watanabe and Chinese doctoral student Cui

Professor Watanabe: Studied at the Graduate School of Environment and Energy Engineering at Waseda University, earning his bachelor's through doctoral degrees, and immediately moved into a post as a supervisor in the research laboratory after graduation. He is a multi-tasker: earlier, he was in the position as the Vice President of the Society of Automotive Engineers of Japan as well as the Director of the Engine System Department of the Japanese Society of Mechanical Engineers. In addition to his scientific research, he was also responsible for giving suggestions on future policy formulation and avenues of the scientific and technological development to the government. He is currently a senior administrative officer of the Central Environment Council of the Ministry of the Environment, a member of the Transportation Policy Council of Ministry of Land, Infrastructure, Transport and Tourism (MLITT), and a member of the Comprehensive Resource Investigation Council of the Ministry of Economy, Trade and Industry (METI).

Cui: Entered Waseda University in 2005 and studied at the Watanabe Laboratory for her bachelor's, master's, and doctoral degrees. Cui entered the laboratory in her junior year to perform some basic research and auxiliary work. In her senior year, Cui joined a project team to do experiments with postgraduates and had a desk and computer in the laboratory.

2 Engineers from Nissan

Professor Kurihara: He graduated in atomistics from Hokkaido University in 1972 and has worked in Nissan's technology research and development department since then. He was engaged in the research and development of electric automobiles for 40 years. In 2008, Mr. Kurihara was employed by Waseda University as a guest professor and began his dual career as an engineer and professor. He spends half of his time in research and development at Nissan, and the other half at Waseda University.

Mr. Yano: He is a department director at the Nissan Research Center. The Research Center consists of five laboratories, four research departments, and one development and testing department. The first research department is

Design Research and Development, the second is Power Service (for developing various power sources), the third is Electric Automobile Development (for developing the whole electric transportation system), and the fourth is Advanced Materials, which focuses on developing new materials like fuel cells. Mr. Yano is the Director of the first research department, the Design Research and Development Department.

4.2 Waseda University and Nissan

Waseda University was the first private university in Japan, and its UIC is always ranked at the top of the list among private universities. "UIC and flexible application of knowledge" is the third of the nine development targets to which the university adheres. The university maintains multiple-collaboration relationships with enterprises in many fields. As most collaborations involve trade secrets that are not to be disclosed, Professor Watanabe advised the author to select Nissan as a case. Nissan is the second-largest automobile company in Japan, and its collaboration with universities is relatively open.

4.2.1 Transition of a Liberal Arts University

The predecessor to Waseda University was Tokyo Senmon Gakko (College), which was not located in such a bustling place. Tokyo Senmon Gakko was a private college with only three majors at the time of its establishment in October 1882. It was renamed as Waseda University in September 1902 after the approval for its reconstruction was obtained. Waseda University, as the best private university in Japan, is a heaven for social elites in Japan. This university has cultivated a number of elite members of media, political, and academic circles, and has made significant contributions to the development not only of academic research and education, but also of the economy.

After 2007, the economic downturn continued in Japan. In addition to the missions of research and education, Waseda University had a new mission: UIC, repaying society with the knowledge of the university. As required by the new era, the potential of UIC for the application of research outputs at this university increased, and the industrial circle showed increasing interest in UIC. Waseda University solved the society's long-term problems through the application of comprehensive knowledge, aiming at promoting technical capacity and economic development in Japan, applying education and research from the university in a flexible manner, and vigorously promoting UIC.

As early as in 1990, Waseda University issued the "Guidance for Conducting Academic Research Collaboration with External Organizations," and established the "Review Council for Conducting Academic Research Collaboration with External Organizations" (generally called the Guidance Council). The Guidance Council was intended to promote UIC at Waseda University based on six principles.[1] In April 1999, the Waseda TLO (a recognized TLO, WTLO for short) was established with the approval of the Ministry of Education and MITI (now METI). It was responsible for the implementation and management of matters related to the university's intellectual property. In 2000, the Intellectual Property Center was set up under WTLO. In December 2001, the Science and Technology Incubation Center (インキュベーション推進室) was established. The main tasks of the center included: (i) cultivating students' awareness and enterprising spirit through multiple ways of supporting teachers, (ii) cultivating excellent entrepreneurs capable of engaging with the international society and contributing to the economic development of Japan, and (iii) creating the new modes of future start-up ventures. In September 2006, WTLO and the Science and Technology Incubation Center were combined (Figure 4.1) to form the UIGC Promotion Center. The major UIGC participants were the Advanced Research Institute for Science and Engineering, the Kagami Memorial Laboratory for Materials Science and Technology, the Global Information and Telecommunication Institute, the Information, Production and Systems Research Center, the Environmental Research Institute, and the Comprehensive Research Organization The UIGC Promotion Center is responsible for the promotion and management of matters related to the collaboration and providing support for these research institutes.

UIGC Promotion Center provides a mechanism for the collaboration between Waseda University and enterprises, and is mainly aimed at: (i) creating and enhancing intellectual property for society, (ii) promoting the development of the local economy and contributing to society via the transfer of technology from intellectual property, and (iii) developing TLO to play the leading role in creating new industries. With its focus on these objectives,

[1]UIGC Promotion Center of Waseda University (2012; http://www.waseda.jp/tlo/jpn/collaboration/principles/index.html). Six principles are: (1) adhere to academic freedom and independence; (2) undertake research that will promote world peace and human well-being and prohibit research concerning military development; (3) enhance the development of research activities and advancement of education at the university; (4) prohibit secret research with research achievement publication restricted (for renewable contracts between research co-signers and co-researchers, the publication date of research achievements is subject to provisions of the contracts); (5) guarantee social justice; and (6) promote collaboration projects based on the principles of information disclosure and procedural democracy. These principles have become the UIC philosophy adhered by Waseda University.

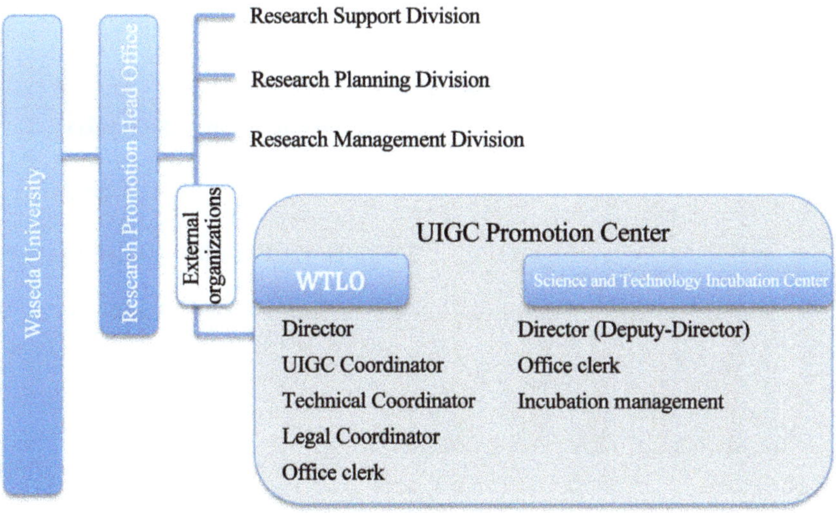

Figure 4.1 UIGC Promotion Center.

Source: Internal information from the UIGC Promotion Center at Waseda University.

UIGC Promotion Center's main tasks consist of the management of intellectual property, technology transfer for the university's science and technology research, and the attraction of more collaborative research projects and contract research projects. UIGC Promotion Center holds three or four science and technology press conferences every year at home and abroad to promote the scientific and technological development and to set up collaboration projects with overseas universities. Between 2005 and 2011, the number of patents applied for by Waseda University in foreign countries went up and down, even showing a decreasing trend, but the number of the patents registered (i.e., the number of patents granted) in both foreign countries and domestically increased consistently. In fact, it increased by a factor of 11.2, and the number of transferred technologies doubled (Figure 4.2).

Since 2002, Waseda University has established close partnerships with Kyushu University, Kyoto University, the Tokyo University of Marine Science and Technology, Tokyo Institute of Technology, Saga University, Kansai University, the University of Tsukuba, Nara Medical University College of Nursing, among others, in succession, and has also established a collaborative relationship with the Tokyo University of Agriculture (a national university corporation), the National Institute of Advanced Industrial Science and Technology (an independent administrative institute), the National

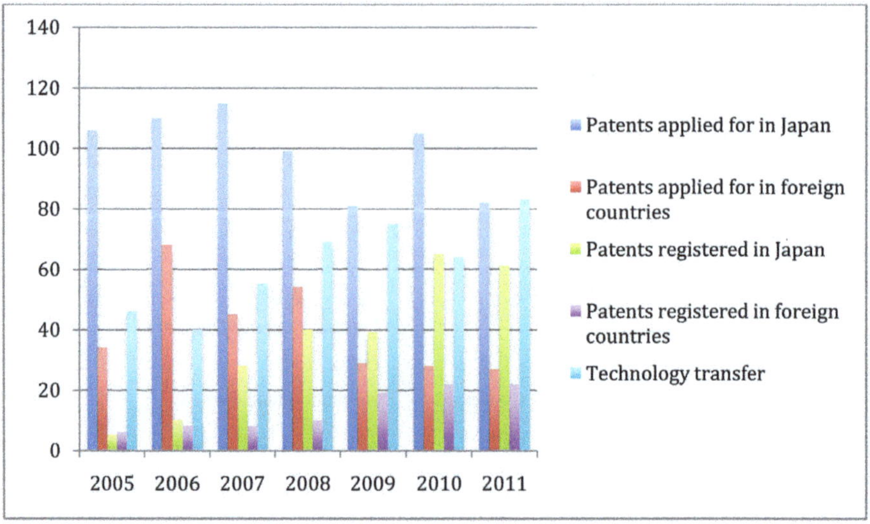

Figure 4.2 The number of patents at Waseda University (2005–2011).

Source: UIGC Promotion Center of Waseda University. http://www.waseda.jp/tlo/jpn/about/data/index .html (2012)

Institute of Informatics (NII), and other research institutions. In addition, it has also established comprehensive collaborative relationships with Adidas Co., Ltd., Mitsui & Co., Ltd., the Japan Bank for International Cooperation, Hitachi Co., Ltd., Nissan Motor Co., Ltd., Nippon Telegraph & Telephone Co., Ltd., etc.

4.2.2 Nissan's Research and Development Difficulties

NISSAN is a Japanese word "日産," which means "Japanese industry," spelled in Roman letters. The company was established in 1911 by Shinjiro Hashimoto and renamed "Nissan" in 1934. Its mission is based on the vision of "enriching people's lives."

Nissan grew increasingly rapidly after 1947. It developed its own products by both introducing and absorbing a large amount of automobile technology from foreign countries, by exporting its products to overseas markets, and by building factories for local production abroad. For more than 40 years following its 1947 boom, Nissan continued its rapid development. As a result, Nissan became one of the largest automobile manufacturers in Japan, second to the Toyota Motor Corporation, and one of the ten largest global automobile manufacturers. After the 1950s, Nissan began to seek support from foreign technology to increase the technical level of its own products. In 1952, Nissan collaborated with the Austin Motor Corporation to develop the Datsun 210, which showed clear improvements in technical specifications. The Datsun 210 won first place in the World Rally Championship (Australia), against fierce competition (Nissan Motor Co., n.d.). Nissan began exporting automobiles to North America at the time of this success. It also launched the Nissan Bluebird, which became a world-famous nameplate.

However, Nissan found itself in a crisis throughout 1990s. Before 1999, Nissan suffered losses for seven consecutive years, with a cumulative total of more than USD 5 billion. This significant loss scared even Ford and Daimler-Mercedes, which had planned to acquire Nissan. Finally, Renault S.A., the biggest motor industry group in France, acquired a 36.8% share of the corporation, constructing the Renault-Nissan Alliance (Nissan Motor Co., n.d.). Renault S.A. rapidly sent its Vice President, Carlos Ghosn, who was known as the "cost killer" and "Mr. Fix It," to hold the post of Chief Operating Officer of Nissan. After Carlos Ghosn's "surgical" reform of the company, Nissan made up its deficits, earned a surplus, and achieved its "recovery plan" within 2 years.

In 1999, Carlos Ghosn became the first person to hold the posts of CEO at both Renault S.A. and Nissan, and English has been the common language at Nissan since then, which facilitated the author's interviews at the company. Some experts predicted that Nissan, whose history at that time spanned more than 70 years, would usher in a "golden age" of development under Ghosn's leadership. According Nissan, its sales volume in 2011 reached 5.76 million cars (Nissan Motor Co., n.d.). Calculated per the consolidated statements of Nissan and Renault S.A., the Renault-Nissan Alliance had become the fourth largest automobile manufacturing group in the world, with its sales dominated by Nissan cars.

Nissan's success is the result of its humanistic, care-based vision. To realize its vision, Nissan used the development concept of the "orchard", including harvest planning, seeding and growth, and soil enrichment (Nissan Motor Co., n.d., para. 2). How to treat technology development and how to formulate strategy and plan accordingly were, Nissan executives believed, the essential issues that enterprises should consider in their development. In a number of cases, the development of the most basic technologies that Nissan expected to launch into the market was the most time-consuming. The "orchard" concept was a panoramic vision that enabled Nissan to consider cooperation in the development of technologies of different types. All activities were carried out using this vision as a starting point, just like the planting and cultivation of different "fruits." Therefore, the entire research and manufacturing process was divided into three stages: (i) the harvest program that prepares the schedule for commercialization of a technology. The researchers would clearly define the value that the technologies could bring to target customers, while the target customers choose to collaborate depending on the performance and function of the technology, including the time schedule. Technology would not be developed for any purpose other than to provide customers with a satisfactory and timely value based on social requirements and market demands. (ii) The seeding and growth stage involves in formulating and implementing plans to realize the "harvest program" before researchers divided the technical factors and come up with a fast and effective implementation strategy. Nissan establishes a collaborative relationship with universities, parts manufacturers, and government officials to form a new management structure that involves in regular progress reporting and consistently introduced improvements in technology. (iii) The soil fertilization phase consists of basic technology development and research, requiring the research team to be able to create a long-term value, such as with technology to increase reliability, which is the soil of the orchard; technology analysis and measurement; and material technology. To increase the efficiency of the process of automobile manufacturing, i.e., technology analysis, measurement and material technology are used to "fertilize" the soil of the orchard and enable human resources and internal processes to play their roles.

Under the guidance of this philosophy, the Nissan Research Center divides the development of automobiles into three stages: research, cutting-edge technology development, and product development. The Research Center consists of five laboratories, four research departments, and one development and testing department. The first research department is Design Research and Development, where Mr. Yano works, the second is Power Services (for developing various power sources), the third is Electric Automobile Development (for developing the whole electric transportation system), and the fourth is Advanced Materials, which focuses on developing new materials like fuel cells. In addition to the five laboratories, Nissan has also established a special organization, the Senior Innovation Researcher (SIR for short), with three senior researchers who are engaged in the world's leading innovative research, to assist the Research Center. The three senior researchers are not on the staff of Nissan, but are professors or engineers retired from the Nissan Research Center who have signed research contracts with the company. If their research plan is approved, Nissan signs individual contracts with them. All the research carried out by these senior researchers is of high-risk, but it is high-return, as well. The SIR system is a very unusual organization in the automobile industry. Established by Nissan in 2012, the SIR system is currently in its initial stages, and functions as a component of Nissan's research strategy. Unique features can be found from its capital to its management. Researchers are permitted to put forward proposals independently, and they have more freedom in the scope of their research and in their methods than the internal researchers. Some seemingly unrealistic proposals may be implemented freely once they are approved.

The Nissan Research Center covers four technical fields: safety, environment, dynamic performance control, and life within the automobile. When researching the design of a new car model, Mr. Yano and other engineers start from these four aspects of the technical design of the vehicle. The research process is generally divided into three parts: social change, technology transfer, and shared innovation. In an ever-changing society, the Nissan Research Center is able to more effectively adapt technology to serve the customers of the future, 10 or 20 years down the road, only by keeping up with changes as they happen on a day-to-day basis. Shared innovation requires the Nissan Research Center to collaborate with universities, enterprises, and the government in order to develop technologies that can compete with those of other motor enterprises. This research consists of two important stages: idea creation and feasibility research. At the idea-creation stage, all parties involved in the project can discuss technology freely, while in the feasibility stage, the researcher can determine the research field and technical targets before finally defining a clear research subject (Figure 4.3). Since staying abreast of societal changes is the purpose of this research in "Social Reform", the Nissan Research Center created a societal research team, responsible for regularly submitting investigation reports to the director of the center. This team investigates changes in Japanese society as well as avenues of development in countries overseas.

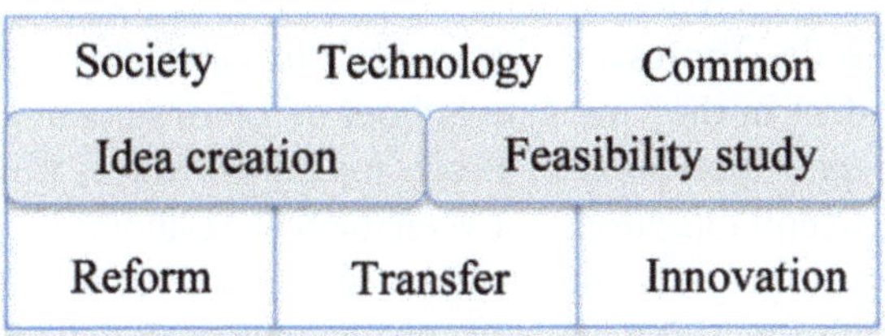

Figure 4.3 Research paths and stages at Nissan (Prepared by the author).

Currently, Nissan has seven R&D departments in Japan. They are the Nissan Research Center, the Hi-tech Center, the Technology Center, and projects in the other four areas. Nissan has also established overseas R&D departments in the US, Britain, and China (Hangzhou). Nissan collaborates not only with domestic and foreign universities, but also with enterprises, governments, and national laboratories. The Japanese government has attached great importance to the development of technology for the future. Currently, one of Japan's national projects is to sponsor Nissan's research on fuel cells. To promote collaboration, the Nissan Research Center establishes overseas research departments in India and the US. These overseas research departments are primarily responsible for the collaborative projects with foreign universities and enterprises by means of contract research and collaborative research. For the contract research, Nissan arranged the objectives, and the universities finished the research projects independently. For collaborative research, both the universities and Nissan participated in the research process; Nissan would sometimes send special researchers to work in university laboratories.

The collaboration between Nissan and Waseda University has a long history, but the signing of a comprehensive memorandum in 2006 strengthened the collaboration between the two institutions. On February 17, 2006, Katsuhiko Shirai, the President of Waseda University, signed the Memorandum of Collaboration with the President of Nissan to carry out far-reaching collaboration in research and development, professional resources exchange, and social contribution. Since the memorandum was signed, the two sides have collaborated by exchanging lecturers, jointly exploring technical instruction, and exchanging professional resources centering on research and development.

4.3 Engine Project

4.3.1 Launch of the Engine Project

As early as the 1970s, Professor Watanabe participated in research for enterprises as a student in the Graduate School of Environment and Energy Engineering under the supervision of Professor Saito, who is now 90 years old. At that time, most collaborations between the Saito Laboratory and Nissan were contract research projects implemented based on the requirements of the enterprise. Before the 1990s, Nissan was economically strong, and had sufficient funds to undertake research that ranged from basic research to product development. Essentially all research was completed in Nissan's own laboratory. The collaboration between Nissan and Waseda University was limited to information exchange and discussion, and the university did not have much influence on the research of enterprises. At that time, Mr. Kurihara conducted research related to the internal combustion engine, and contact with the Saito Laboratory only concerned information exchange for the improvement of technology.

Broad collaboration between Nissan and universities actually began since the 1990s. When Professor Watanabe succeeded Professor Saito, collaboration between the laboratory and Nissan became more diversified. Professor Watanabe divided the research in the laboratory into three types. The first was internal research, i.e, independent scientific research at the university, focusing on in-depth basic research. The second was contract research. This was special research required by the automobile industry, and its results would be applied to the development and manufacturing of new automotive products. The third was collaborative research, including lectures, forums, information exchange conferences, consulting, etc. Collaborative research could be initiated by universities or through collaboration with enterprises to apply for scientific research funds from the government. For collaborative research, Professor Watanabe could select an appropriate collaborative party out of the fifty-four members of the "Waseda Dynamic Research Agency" that he had himself established. If the research proposal was approved by the selected enterprise, Professor Watanabe and the enterprise could jointly apply for scientific research funds from the government. Professor Watanabe further explained the difference between contract research and collaborative research this way: generally, contract research concerned the special research projects of enterprises. For instance, Nissan's engine project primarily involved research on the combustion, simulation, and processing systems of the engine. It was generally the case in a contract research between a university and an enterprise that the university would conduct relevant research on a single subject independently. However, the collaborative research had a larger scale. It involved at least one university and more than one enterprise (Table 4.1). Professor Watanabe and other professors called it "pre-competition research," signifying that it was not competitive at the time, but would be in the next 10 or 20 years. The enterprises in the collaboration would share the competitive research achievements, and the collaborative activities were wider, deeper, and more comprehensive.

Table 4.1 Similarities and differences between contract research and collaborative research

Name/Similarities and Differences	Collaborative Party			Subject Nature	Source of Funds
Contract research	One university	One enterprise	Government	Currently competitive	Government or enterprise
Collaborative research	Several universities	Several enterprises	Government	Prospectively competitive	Government or enterprise

Source: Prepared based on exchanges with Professor Watabe.

The success of research on fuel cells and the mating engine at Nissan was the result of a joint effort by Nissan, Waseda University, and many other research institutions. Over the course of this research, Mr. Kurihara was the key figure in the collaboration between Nissan and Waseda University. The electric automobile was Mr. Kurihara's first subject after he joined Nissan in 1972, and, unexpectedly, it became his life-long research subject. At that time, the technology of the electric automobile was at the preliminary stage. The fuel cell was huge, heavy, and expensive. In Mr. Kurihara's tests, the output power of the fuel cell was only 0.5W/km, and 80W/km was required to run an automobile. The research team led by Mr. Kurihara had to increase the power of the fuel cell as well as reduce its weight and volume. Eventually, given that the fuel cell could not actually be installed in cars, the Nissan Research Center ended the Kurihara team's research.

In 1984, Mr. Kurihara returned to research on the traditional internal combustion engine. His research used low-carbon, environmentally friendly methanol gasoline, and thus was still related to the development of green cars. At that time, the development team led by Mr. Kurihara often exchanged technology and shared information with the Watanabe Laboratory. This was an early stage of the eventual collaboration between Nissan and Waseda University. Mr. Kurihara returned to fuel-cell development in 1998, when laboratories in other countries began to make great progress in their research in that field. In 2001, Nissan sent Mr. Kurihara to the US to build a research and development laboratory, the first research and development center built by Nissan in a foreign country. Mr. Kurihara's task was to continue developing a fuel cell that could be applied installed in an electric automobile. At that time, fuel cells were widely used in the aerospace field. After the successful application of fuel cells in the Apollo spacecraft in 1960, the hydrogen fuel cell was widely used in the astronavigation field. At that time, the United Technologies Corporation (UTC) was a famous laboratory for aviation technology research with rich experience in researching the fuel cell.[2(UTC)] In the US, Mr. Kurihara collaborated with UTC to master the technology of the fuel cell. In 2004, equipped with significant experience in research and development of fuel cells in the aerospace field, Mr. Kurihara returned to Nissan headquarters in Yokohama. Mr. Kurihara devoted himself to the

[2]United Technologies Corporation, formerly known as the United Aircraft Corporation, is engaged in aircarft manufacturing and air transportation. The United Aircraft Corporation, the Boeing Company, and United Airlines were formed in 1934 due to the American government's refusal to continue to allow the merging of airline companies with aircraft equipment manufacturing companies.

application of this research to the development of fuel cells in cars, and to researching and manufacturing mating engines and fuel cells at Waseda University and other universities and research institutions. Finally, Nissan launched its Leaf, the first electric car on the international market, in 2010. At the 2010 New York International Auto Show, the Leaf defeated the BMW 5 Series, the Audi A8, and other heavyweight nominees to win the title of "2011 World Car of the Year." Currently, the Leaf is sold in thirty countries, including the US, Germany, France, and Portugal. Excluding the period when the research was suspended, 20 years were spent on researching the fuel cell used in the Leaf. The launch of the Leaf was the result of Nissan's "pre-competition research" strategy, as well as their joint efforts with Waseda University and other research institutions.

Currently, the collaboration between the Watanabe Laboratory and Nissan is considered collaborative research, which is based on constant exchange visits and information exchange between universities and enterprises and is accepted and supported by the government. Waseda University employs Nissan engineers as guest professors or part-time professors who spend half of their time at Nissan and the other half at the university and are paid both by both, with each party responsible for 50% of their pay. The 50% of the salary paid by the university generally comes from government project funds granted for collaborative research. These government project funds cannot be used to pay the salaries of university professors; they can only be used to pay the salaries of guest professors sent by enterprises. Enterprises can reduce their original expenditures by half, while universities do not need to bear any costs. As a result, both sides are satisfied with this system. Professor Watanabe sustained his laboratory by applying for various kinds of government project funds. Kurihara, meanwhile, has continued the collaboration between Nissan and Waseda University in a new way since 2008. Invited by Professor Watanabe, Kurihara became a guest professor at Waseda University and is engaged in research and teaching in the Graduate School of Environment and Energy Engineering. Currently, he is primarily responsible for two research projects: research on mobile power, which is related to his primary research interest of electric automobiles and buses, and research on the smart grid, which is aimed at realizing safe use of the grid with advanced equipment and technology. His goal is to build a smart grid for a residential community network that provides a platform for effective interaction between electric automobiles and environmentally friendly communities. Currently, Professor Kurihara teaches a graduate-level course once a week at Waseda University in addition to his participation in research projects sponsored by the government.

In addition to Nissan, Watanabe Laboratory has also built collaborative relationships with Honda, Toyota, Suzuki, Mazda, and other enterprises in the automobile industry. In this laboratory, students are able to join different project teams to carry out research on different subjects. Watanabe Laboratory collaborates not only with enterprises, but also with national laboratories, such as the National Traffic Safety and Environment Laboratory (NTSEL), the Japanese Automobile Research Institute (JARI), and the National Institute of Advanced Industrial Science and Technology (AIST).

After the Leaf conquered the international market, Nissan began to collaborate with Chinese universities. In August 2012, Professor Kurihara was invited by China's Tsinghua University to give a lecture on the current state of electric vehicle technology. At this lecture, about ninety students from various universities gathered together and had an energetic discussion, impressing Professor Kurihara. Professor Kurihara believed that green and environmentally friendly technology would inevitably bring about a new era in the global automobile industry.

4.3.2 Collaborative Mechanisms

In the collaboration process between Watanabe Laboratory and Nissan, researcher exchange, task division, and technology selection ensure a smooth collaboration from subject selection for research projects and organizational management to the achievement of the objectives.

4.3.2.1 Researcher exchange

The success of collaboration between a university and an enterprise depends on mutual understanding. In collaborative research, both sides usually exchange researchers in order to gain a precise understanding of the research process, thereby enabling them both to meet common goals.

Professor Watanabe mentioned the importance of the enterprise doctoral course of Waseda University to collaboration:

> We have two visiting researchers in our laboratory from the enterprise. They often come to the laboratory to discuss research progress and make the next research plan with us. Staff of the enterprise can also study at Waseda University for doctoral degrees and their tuition is usually paid by the enterprise. They are not required to complete the courses of study, but they can obtain the doctoral degree only if their doctoral thesis meets the review requirements

> and they give an oral defense. The enterprise's doctoral candidates shall come to the university at least once a week to participate in collaborative research. University professors will strictly assess the theses submitted by staff from enterprises as required. Generally, a doctoral thesis shall be modified several times before being qualified. In addition to the doctoral thesis, staff from enterprises shall publicly publish four to five articles before obtaining the degree. With professional training at the university, the enterprise's doctoral candidates will promote the development of basic research, which is weak in enterprises.

It was Professor Kurihara, with his dual experience, who was able to truly understand the essence of research from the perspective of both the university and the enterprise, and, therefore, able to promote the organic connection of the two sides to realize a mutual benefit.

4.3.2.2 Effective task division

Over the course of more than 40 years of his research at Nissan, Professor Kurihara participated not only in collaborations with Waseda University, Keio University, Hiroshima University, the University of Tokyo, etc., but also in various collaborative research projects with the government, enterprises, and national laboratories, as well as overseas universities. As he noted,

> The collaborative research between universities and enterprises can be divided into three stages: research stage, cutting-edge engineering development stage, and product engineering development stage (Figure 4.4). In collaborative research between universities and enterprises, the universities mainly undertake the majority of basic research and other research work. In terms of results and benefits, this research involves a high degree of risk. Generally, basic research means research on mathematics, physics, chemistry, and other basic subjects, but the research that we refer to is basic research in a special field. The tasks of universities and enterprises are different in collaboration. Generally, the universities undertake research that involves a higher degree of risk, while the enterprises focus on technology development and product manufacturing, which consume the greater part of the funds. Enterprises only spend a small proportion of funds on basic research.

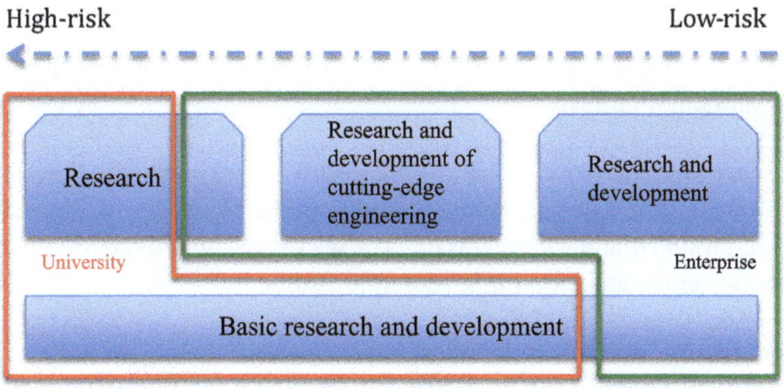

Figure 4.4 Stages of UIC according to Professor Kurihara.

Before the 1990s, Nissan had abundant financial resources, and was able to independently afford basic research. But due to the bursting of the economic bubble, Nissan and other large enterprises were increasingly unable to independently afford this kind of research. As a result, collaborative relationships between universities and enterprises began to change. In the opinion of Professor Kurihara, the collaborative relationship between Waseda University and Nissan entered its "transition period" at that time, and changed from the old collaborative mode of information exchange to a new mode of task division. Currently, collaborative research between Nissan and Waseda University operates in a specific mode of task division. As shown in Figure 3.4, Waseda University was mainly responsible for (basic) research, while Nissan provided cutting-edge engineering and product engineering.

Mr. Yano, the Director of the Nissan Research Center, agreed with the above description of the division of labor:

> Nissan Research Center divides the development of automobiles into three stages: research, cutting-edge technology, and engineering development and product development. In the product development stage, the vehicle body is manufactured, while in the cutting-edge technology development stage, high and new technologies for fuel cells are developed. Research is the first stage, and is the main task of the Nissan Research Center. The three stages are of the same importance. The research stage is to prepare the technology for the future 10–20 years, while the product development stage is the basis for Nissan to develop and grow. Our technical center is responsible for the later product development and research, and the

> high-tech center for research on high and new technologies. It is very necessary to clearly define the tasks of the Nissan Research Center and universities.

Cui, a doctoral student from China, participated in various collaboration projects with Professor Kurihara while studying in the Watanabe Laboratory for 5 years, and found that the demands enterprises placed on universities were in fact a kind of dependence on basic research results:

> The laboratory mainly undertakes basic research and generally does not participate in the research and development of products for enterprises. Basic research requires considerably more human and material resources and time, which can be provided easily by universities. The enterprise, on the other hand, looks to manufacture profitable commercial products. Enterprises will not put much money into unprofitable and expensive basic research. Even if the enterprises invest the same capital in basic research (as universities do), they may not ultimately obtain the expected results. But enterprises can invest *some* money in universities and gain a much higher return. All the projects we participate in involve basic research and application research. The product development is later undertaken by enterprises. Of course, enterprises will consult the university when they have problems during the product development.

Therefore, as a result of their inability to undertake basic research, enterprises rely on universities. Professor Watanabe's viewpoint further confirmed the necessity of task division between universities and enterprises: "As to collaboration with universities, enterprises have two specific targets: obtaining excellent potential staff from universities and making use of abundant basic research achievements in universities."

The university researchers, working within a supportive culture, made a much more substantial contribution to the performance of UIC. A continuous partnership was favorable for greatly reducing the transaction costs to both sides of the UIC (Tornatzky & Bauman, 1997).

4.3.2.3 Technology selection guidance

Currently, collaboration between Nissan and universities is quite broad in scope. Nissan not only collaborates with universities in Japan, but has also established relationships with universities, in the US, Europe, China, India among others. Nissan also cooperates with enterprises in more than twenty countries, including China, North America, the United Kingdom, and Spain. Its overseas technical centers are mainly in the US, Europe, and China.

In selecting the universities to collaborate with, Professor Kurihara believed that technology is the priority:

> Where there is more advanced technology, there is more collaboration, whether in overseas or domestic universities, be it enterprise collaboration or university collaboration. Modern automobiles require video cameras and radar to monitor driving safety. Some universities in the US have ICT technology and smart analysis technology to monitor driving safety, while universities in Japan have advantages in ACM cells and semiconductor technology. In addition, Japan leads the world in research on materials. For example, Tohoku University maintains a high level of new material development technology.

With regard to take the technology as a collaborative objective, Professor Kurihara recalled a collaboration project with a university in the US from 1994 to 1998. In 1994, he stayed in California and was mainly responsible for researching the rotor system of engines and designing and manufacturing the rotor, and internal combustion engine. At that time,

> I discussed the research with several professors in California. They were all very communicative. Moreover, Californian people were hospitable and friendly. The kind and easy-going talks with American professors deepened the discussion on development of the green engine. At that time, the environmental quality standards of California led the world. The California Air Resources Board (CARB for short) was the government organization responsible for environmental management. It created the world-leading air clearness standard, which was followed by the rest of the US. Later, Europe and Japan applied the American air clearness standard. To produce an environmentally friendly automobile that complied with the air clearness standard, we specially established a connection with CARB. At that time, I was very surprised that the officers of CARB, the government organization that formulated the world-leading standard, talked with me warmly. I did not expect that I could have communicated with the government officers so easily. Moreover, they even expressed that they would consider appropriate

> modification to the standard when they believed my viewpoint was reasonable. It was impossible in Japan at that time. The social structure of Japan was from top to bottom. The government was always in a high rank and it was very hard for common people from enterprises to communicate with government officials. The government would not change the standard once it was formulated even if some views provided by people from enterprises were better.

Even though 20 years had passed, he was still in high spirits when talking about this. It is thus clear how excited the young engineer was when he found his opinions might bring about changes to the standards formulated in California. Of course, the Japanese government was changing as well. Professor Kurihara remarked that he feels that it has been easier to communicate with government departments in the last 10–15 years. His opinions have not always been directly rejected by the government, and in some cases have even been accepted.

The development strategy of taking technology as a wind vane has clearly defined the collaborative targets of enterprises, which led to the establishment of clear and effective intentions for collaboration with universities, providing a strong foundation for successful collaborations.

4.4 Win–Win and Conflicts

4.4.1 Circumvention of High-risk Enterprise Input

Large enterprises in Japan have always insisted on an independent "research and development at their own expense" system, but with the acceleration of the globalization process and a shorter product research and development cycle, large enterprises have had increasing difficulty maintaining such a system. Professor Kurihara ruefully explained the advantages and disadvantages of the traditional system:

> In the past, enterprises carried out research and development by themselves. They had to be engaged in many fields, such as fuel cells, electric automobiles, battery, and engines, so as to have some understanding of all these fields. But involving too many fields was not favorable for successful development and application, and moreover, my individual ability was limited. Now we are able to collect technology and thoughts from famous professors in various fields. Integration of advantages from various experts is far more advantageous than the independent development of Nissan.

With the slowdown of the markets and the problems of the products in the 1990s, it was very hard for Nissan to independently undertake high-risk basic research and technology development. Professor Kurihara said that, based on his personal experience,

> The high-risk research and development, for instance, on the fuel cell, could not be independently finished by the enterprise any more. Challenges from reality forced us to turn to universities to seek cooperation with professors who could undertake high-risk research and development. Finally, we began collaborative research with professors all over the world. By searching for researchers with advanced thoughts and frontier concepts all over the world, the vision of research of Nissan under "research and development at its own expense" was greatly broadened.

After researching the development of the electric automobile for such a long time, Professor Kurihara realized that the scientific research was different than product development. Product development can obtain successful results in the short term, while the basic research for supporting new high technology requires a growth cycle, and may eventually be fruitful after several "intermissions" (Figure 4.5). Consequently, the basic research independently undertaken by the enterprise involves a higher degree of risk.

The launch of the Leaf in 2010 brought a high capital return to Nissan. The SIR Team, which employs senior researchers from enterprises and universities for world-leading creative research, was established under the auspices of

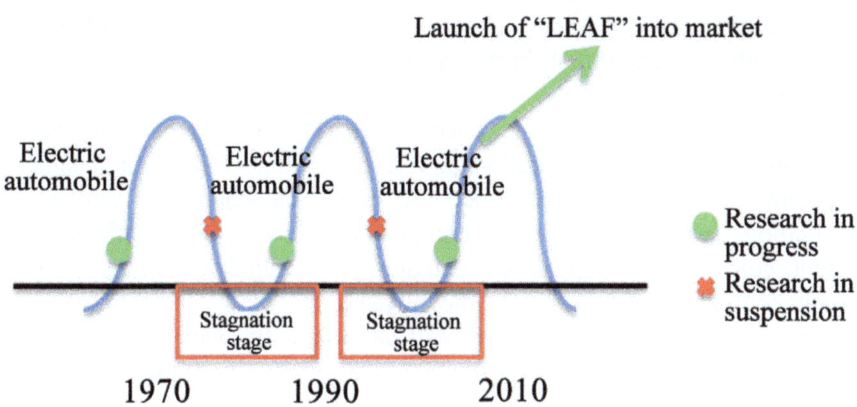

Figure 4.5 Research and development cycle.

the Nissan Research Center in 2012. Incorporating high-risk technology development into internal research and development systems once more indicated that Nissan was paying close attention to basic research. Nissan's collaborative development with universities further showed its dependence on basic research achievements.

4.4.2 Enterprises' Social Concern Aroused by Universities

Nissan's development vision is to "enrich people's lives." These simple words embody the development concept and goals of the enterprise. Professor Kurihara explained the real meaning of this vision:

> We aim at providing: first, high-quality automobiles; second, life within the automobile; and third, community life. Producing high-quality automobiles is the basic goal of motor enterprises. Enabling people to enjoy the fun of driving and traveling with family is the life Nissan attempts to provide. The environmentally friendly automobile will form a brand new community life and facilitate people's daily lives. The high-quality automobile is the hard power for the development of Nissan, while automobile life and community life constitute its necessary soft power.

Thus only by basing development on the predicted society of the future can enterprises keep developing in perpetuity. Professor Kurihara was full of enthusiasm for the research he has undertook at the university because in the past he had only engaged in technology research, whereas his present research enables him to combine new technology with the construction of future communities. He was excited about improving society with technology. Professor Kurihara insisted that, in the integration of various types of future construction.

The automobile industry shall expand collaboration with other fields. Now Nissan has begun collaboration in ICT technology for research and development of the smart automobile. The newly launched Leaf applies ICT technology, which is a piece of complicated network communication equipment. Voice communication between driver and vehicle can be realized with such equipment. For instance, the owner can "send" an instruction before going to work, and the vehicle will automatically wait at the door. The integration between electric automobile and community requires cyclic utilization of energy. In traditional design, solar panels may be installed on the roof to absorb the solar energy in the daytime and store it as electricity, but the equipment for energy storage is very expensive. If the electric automobile is selected as equipment for energy storage, the costs will be reduced greatly. The energy stored by the electric automobile in the daytime can be used at night.

Professor Kurihara was proud of the Leaf's fuel cell. With an economical and efficient battery, the electric automobile has become the favored alternative to expensive energy storage equipment. As a large enterprise, Nissan is very concerned about society, and this concern was constantly strengthened during collaboration with universities. As a result, the company's social responsibility was enhanced. Enterprises improve their technology by focusing on society. Attention to the future of society may become the necessary soft power for the development of enterprises.

4.4.3 Enterprise's Development Value Constantly Affected by the Ideas of the University

Is it enough to only rely on profit growth to develop enterprises? What kind of rare energy can enterprises get from universities? After years of conflicts, Professor Kurihara finally concluded,

> The most important thing that the universities provide for enterprises is ideas. I once participated in a collaborative research project on biological diversity. During the collaboration, Stanford University introduced the concept of the importance of biological diversity to the development of the world to enterprises for the first time. Universities focus on the future development and living status of human beings. It is very important to spread such ideas in enterprises to help enterprises realize the impact of industrial development on the future of human beings with the help of these ideas. As part of social development, enterprises can make a contribution to human beings through their own efforts. But universities shall play their role in conceptual introduction.

Professor Kurihara considered this as a very successful and pleasant collaboration. In the collaboration with Stanford University and some non-governmental organizations, various parties often convened meetings to discuss solutions within the industrial circle. They tried to imbue automakers with a sense of responsibility for the protection of biological diversity and discussed what enterprises could do about it. Through this collaboration,

Nissan learned about the relevant environmental background and could consider issues such as how to respond to global warming, air and soil pollution, and biological diversity. Professor Kurihara believed that "Enterprises will have a clear understanding of their environmental background and be able to deal with the impact imposed by their own industries on the environment. This is very important to not only Nissan but also the whole automobile industry. Although it brings no direct benefits to enterprises, it is necessary."

Another collaboration that impressed Professor Kurihara was the collaboration with the World Business Council for Sustainable Development (WBCSD). WBCSD is an association of some two hundred international companies that are committed to a sustainable development model for economic growth, ecological balance, and social progress. Nissan, Toyota, and Honda, are among its members. Nissan undertook a sustainable mobility project, and is responsible for researching how to develop an automobile with sustainable mobility, how to dispose of and reduce carbon dioxide in exhaust, and how to develop the technology required for an electric automobile or fuel cell. Nissan will endeavor to realize the sustainable mobility society by 2050. Professor Kurihara and lots of other engineers have devoted themselves to research on reducing exhaust pollution, but with the increase in vehicles used every year, environmental pollution has not really been mitigated. Therefore, the development of the green energy automobile has always been Nissan's primary development objective, and the development direction insisted on by Professor Kurihara.

Although they were willing to make efforts to achieve various goals, fierce market competition has kept enterprises too busy to consider the progress of society and the future of human beings. Therefore, the conceptual introduction of the university was important. Professor Kurihara explained,

> The operation of enterprises is too practical, driven by profit and guided by the market. Nissan is mainly concerned with gaining profits, but it is not enough to gain profit only, so the company is more than willing to promote and create a more beautiful world. Nissan's awareness of the world and concern for the future propel Nissan to consider more about its development direction and what kind of products should be presented to the next generation. The future of society has become an important focus for Nissan. In development, enterprises may be so busy pursuing "profits" that they forget "concepts." Exchange with universities helps enterprises get the lost "concepts" back.

Is it necessary to turn universities into market-led organizations? The pursuit of various objectives by enterprises seemingly answers the question of many professors, which has existed since the commencement of UIC. According to the finding of this case, the answer is no. Universities are places where philosophical questions can be answered which may guide industries directions; therefore, enterprises preferred to collaborate with them instead of other industries to get their lost "ambitions" back.

4.4.4 Compensation for the Lack of Hardware and Software for Scientific Research in Universities

Research universities' strength is scientific research. Through collaboration, research universities and enterprises can optimize and integrate each other's resource advantages. Professor Watanabe often sends his students to large national laboratories, such as the National Institute of Advanced Industrial Science and Technology (AIST) to participate in collaborative research. AIST was the largest research institution in Japan at the time. It was established through a joint effort of fifteen national research institutes under the Industrial Technology Institute of the former MITI. As the largest technology research institution, its research focuses on biotechnology, chemistry, electrics, geology, information technology, mechanics, materials, metrology, and other fields. AIST emphasizes the application and commercialization of pure scientific discovery, as well as interdisciplinary research and development. Professor Watanabe said, "These large national laboratories have expensive and advanced experimental equipment that complements the equipment in universities. Students can make full use of the advanced equipment in national laboratories through collaborative research."

The same can be said of collaborations with enterprises. Research in universities is generally carried out on a particular part of the experimental equipment, but enterprises have large, comprehensive testing equipment. The doctoral student, Cui, had strong feelings about this: "Through collaboration, we are able to get a clear understanding of the role of our research field in the actual operation of a complete vehicle, thus getting better at coming up with response strategies for the next research step."

The collaboration between universities and enterprises not only makes up for the lack of experimental equipment and materials in universities, but also cultivates students' critical thinking ability on actual problems through exchange with engineers from enterprises. As a guest professor, Professor Kurihara said, "I can give classes to students on research and development in a motor enterprise, but the university professors cannot." Professor Kurihara participated in the research and development of the Leaf, so he can tell students about the research and development process for the most advanced electric automobile.

In collaborations with enterprises and national laboratories, universities make better use of the advantages of the other side, and further promote their development through optimization and integration of resources.

4.4.5 Assistance in Cultivation of Doctoral Students and Employment

Seeing the advantages brought to universities by collaborative research, Professor Watanabe began to participate in collaborative research projects more actively. Almost all the students at the Watanabe Laboratory participated in at least one collaborative research project with the automobile industry. He noted,

> When the enterprise sends engineers to the university for discussion meetings, all the students in relevant projects will attend the meeting, and they understand the demands of the enterprise through communication in the meeting. Such practical collaborative research effectively stimulates students' enthusiasm for their study. Students always work with staff from enterprises energetically, and finally find a reasonable foothold for theory in practice.

In the view of Professor Watanabe, students not only benefited from collaboration in scientific research, but also gained opportunities for future employment:

> All students in our laboratory have the chance to participate in collaborations with Toyota, Nissan, Suzuki, and other motor enterprises. The laboratory also keeps a close relationship with Honda. About five to six students from the laboratory are employed by Honda and about ten by Nissan every year. In addition to automobile enterprises, students also participate in collaborations with Komatsu. Komatsu is one of the biggest manufacturers of engineering machinery and mining machinery in the world. . . . This is a contribution to universities made by the enterprises, or a mutual contribution to each other. Enterprises provide a lot of chances for students to participate in practical study, and then the students can work better for the enterprises. It is a virtuous social circle.

These collaborations help students understand the scientific research demands and operation modes of enterprises in detail. Meanwhile, the number of theses and patents has also increased. Naturally, the attraction of universities has increased.

4.4.6 Distance between Ambitions of Universities and Reality of Enterprises

It is the first step, which is the most troublesome. Both the negotiation before the decision to begin collaboration between a university and an enterprise, and the process of setting up a project at the beginning of the collaboration, are full of difficulties in understanding and discussion. In his more than 40 years of research and development work at Nissan, Professor Kurihara collaborated with many universities; having been through successes as well as failures, he found that, in some cases, it was only through a joint effort that both sides were able to overcome their difficulties.

> The first problem to be solved is to find an appropriate university. University professors seek truth through scientific research, and have a strong sense of curiosity. But enterprises need to discover the technology society and human beings require. Sometimes, professors' ideas are too idealistic to be realized, and are therefore not useful to enterprises. As a result, enterprises must look for universities that suit their priorities.

Research and development in enterprises is different from the research conducted by universities. At the beginning of a collaboration, it is very hard for the staff from enterprises to understand the professors' point of view. Through repeated discussions, the two collaborative sides gradually come to accept and understand each other. Although the enterprises do not understand the ideas as much as the universities, the distance between the two sides is shortened. Therefore, mutual understanding and the establishment of common targets are the important factors for the beginning of an effective collaboration.

Collaboration is a way to facilitate technology innovation and the development of enterprises. According to Professor Kurihara's experiences, it is hard to find a good research direction. In most cases,

> "a research subject that a university professor may consider good may not be required in manufacturing of the desired automobile. We are not sure which advanced technology is required in the

> automobile market. Sometimes the technology, which we think is important, may not be needed by the market or society."

Discovering market demands and technology development directions with university professors is a target that Professor Kurihara and other engineers continue to work toward. To this end, they are continually seeking good communication and management systems. In the past, enterprises only needed to develop hardware, but now they have to develop both the hardware and software, as well as the technology required by society. Research and development that integrates hardware, software, technology, and society makes the organization and management processes more difficult. Professor Kurihara called this period a "transition period," i.e., the transition from a past single-technology development model to the current comprehensive technology development landscape. The "transition period" is something enterprises expect to experience, but development during the transition period is still difficult.

4.4.7 Ownership of Research Achievements

Sharing research achievements is often at the root of conflicts between universities and enterprises. Sometimes, motor enterprises are not willing to publish the data from research achievements, but Professor Watanabe and other professors expect to publish the data in public academic conferences. Sometimes, Professor Watanabe had to wait 2–3 years after negotiations with the enterprises before being able to publish his work. If the research achievements are published in a thesis, they cannot be used for applying for patents. As a result, patent application generally is conducted first, but Professor Watanabe did not consider this a big problem.

> Because universities and enterprises must reach an agreement on the terms of their collaboration, certain issues can easily provoke conflict. Some enterprises require cosigning published research, while some do not want the public to know they have collaborated with Waseda University. Generally, anonymous collaboration is reserved for research in a specialized field or with a high level of competition.

The way universities present research achievements depends on the varying requirements of enterprises. Universities have no voice in this area. To avoid conflict in the sharing of later research achievements, university professors and enterprises must reach an agreement on handling expected research achievements and then sign a formal contract. Consequently, to effectively guarantee the rights and interests of the two collaborative sides, universities will generally exercise the right to express their opinions in the preparation stage before collaborative research, and after that will only be able to perform under the stipulations of their contracts.

4.5 Summary

Waseda University, the best private university in Japan, has also held an advantageous leading position in UIC. Because the university had to compete for research funds against national universities with many advantages after national university corporatization, it became more active in UIC.

The UIC center, or the UIGC Promotion Center, of Waseda University is much simpler in structure than that at the University of Tokyo or Hiroshima University. It consists of only two departments: WTLO and the Science and Technology Incubation Center. Reference Case III is a UIC project in which Mr. Oota, the principal of the Science and Technology Incubation Center, participated. This project is a significant collaboration project involving the university, an enterprise, and the government. Still, it is not clear that the objective of the UIC promotion organization of Waseda University is to participate in large collaboration projects with multiple players.

The cases selected in this chapter are typical projects in engineering research subjects. Professor Watanabe is very active in UIC as he is holding multiple jobs. The author found that the Waseda case was the most successful case among the cases at the University of Tokyo, Waseda University, and Hiroshima University. In the collaboration studied, both the university and the enterprise not only pursued the technical functions of their innovations, but also began to focus on the broader, more social significance of their research and development.

5

Collaboration between Hiroshima University and Mazda Motor Corporation

CX-5 Internal Combustion Engine Project of the Graduate School of Engineering

5.1 Introduction

The Mazda Motor Corporation (hereafter referred to as Mazda), known as the largest motor enterprise in the Chugoku/Shikoku area, is the UIC partner of Hiroshima University with the wide-ranging and long-running collaboration. Therefore, when the author was seeking a case in the engineering field at Hiroshima University, Mr. Matsui's first recommendation was the collaboration between the Noda Laboratory and Mazda.

It took the author about a month after establishing contact to finalize the schedule of interviews and observations. During this period, it was Mr. Matsui who acted as an intermediary between Mazda and the author. Arai, who was directly responsible for the collaboration with Mazda, was supposed to accompany the author to visit Mazda, but he was not confident in his English, so Mr. Matsui was invited instead.

At 8:30 am on November 7, 2012, the author met Mr. Matsui at the Saijo Station in Higashihiroshima and embarked on the visit to Mazda's Mukainada Headquarters. The author had always believed that it was not wise to visit an enterprise with a university professor, but chose to do so in this case because of Mr. Matsui's enthusiasm and her own language difficulties. She anticipated that his presence would result in some conflicts. It later transpired that this manner of interview would bring unexpected results. It was possible to avoid some conflicts by applying skills that will be discussed in the research introspection section in detail.

Mr. Kawakami (川上), the Project Engineer, welcomed the author and her escort warmly. After several complicated security procedures, they came to Mazda's meeting room. Visitors are required not only to be registered with

Mazda, but also to keep their mobile phones and cameras in a locker outside the office area. The author was fortunate enough to be given permission to take her mobile phone and camera into the office area for sound recording and to take photos, while Mr. Matsui locked his mobile phone in the locker as required. The author found out later that all Mazda staff are required to lock their mobile phones in their private lockers during office hours, and then punch their cards to get into the internal office area. The author and Mr. Matsui had temporary cards for entering the office area, which they returned afterwards. Before entering the office area, Mr. Matsui asked Kawakami if it would be possible to visit the Mazda Museum after the interview. Kawakami specially arranged a visit to the Mazda Museum so that it would be possible even if the number of appointments to visit the museum, which were limited in number each day, reached the upper limit. After finishing the interviews in the morning, the author and Mr. Matsui visited the Mazda Museum together in the afternoon.

The interview with Kawakami was a part of the research plan, but meeting with Director Nagashima was really a coincidence. Yukio Hosaka was the engineer that the author had become acquainted with during her visit to the Satake Corp. with Mr. Matsui. During a brief chat, he had said he could introduce the Director of the Mazda Technology Research Institute (his good friend) to the author during her visit to Mazda. She had planned to visit the Director in the middle of December, but as the schedules of all three parties conflicted, this visit instead took place at the beginning of the following year. On January 18, 2013, Mr. Hosaka accompanied the author to the Mazda Technology Research Institute in Hiroshima. The author began another interview with influencing factors, but this time, the influencing factors were from the same perspective. Maybe because Mr. Hosaka and Mr. Nagashima were well acquainted with each other, the interview proceeded with in-depth, detailed discussion. Finally, Mr. Nagashima even showed the author the future development plan for the collaborations between Mazda and various universities. The author was astounded by Mazda's detailed UIC development plan. But as Mr. Nagashima said, this plan involved many commercial secrets, and as a result the author was not able to make a copy or take a photo. The author was nevertheless very impressed by the fact that Mazda paid such great attention to research in universities and took an active part in UIC, which was reflected in the development plan for future collaboration with universities.

People involved in the collaboration project are briefly introduced below:

1 Professor Noda and his students

Professor Noda: Studied mechanical engineering from his undergraduate years through his doctoral degree. He was at Hiroshima University, where he then became a Professor of Engineering. He studied with Professor Hiroyuki Hiroyasu, starting in 1978 and worked with him for more than 20 years. Eventually, Professor Noda succeeded Professor Hiroyasu at Hiroshima University. In the 1970s and 1980s, Professor Hiroyasu was famous in the engine-manufacturing field across Japan and the world for his research on jets. Professor Hiroyasu started the university's collaboration with Mazda, and Professor Noda maintained and expanded it. At present, the Noda Laboratory maintains collaborations with two enterprises, Mazda and Isuzu. Two of the eighteen students in the laboratory are engaged in the Isuzu project, while the others work on the Mazda project.

Li: A Chinese student. He graduated from Dalian University of Technology and enrolled at Hiroshima University in October 2011 to pursue his doctoral degree. His doctoral degree was sponsored by a MEXT scholarship. According to the requirements of this scholarship program, he needed to gain practical experience. Li had a 2-month internship at Mazda, arranged by Professor Noda, and then finished the practice report to submit to MEXT in December. Li was on the diesel spray research team, and directly participated in Mazda's collaboration project after his practice. There are generally seven to eight students in the diesel collaboration project, of whom about four are engineers from Mazda.

Chen: A Chinese student, received his bachelor's degree from Nanchang University and master's degree from South China University of Technology. After completing his master's degree, he went to Hiroshima University to begin his doctoral degree in October 2011 through a program sponsored by the China Scholarship Council. As part of the collaboration project with Mazda, Chen participated in the research of the gasoline spray team, and was mainly responsible for numerical simulation. Chen had not studied Japanese before arriving in Japan. He found that studying in Japanese increased the difficulty of his research, and so chose to complete his thesis in English. Due to language difficulties, Chen communicated mainly with Professor Noda during his research.

2 Engineers from Mazda

Kawakami: Has worked at Mazda as an engineer since his graduation from Tohoku University in 1986. He was only aware of the history of collaboration between universities and Mazda after 1986, but the author found some inaccurate chronological information provided by Mr. Kawakami according to the information obtained in the interview with Nagashima, the Director of Mazda Research Institute, which will be discussed in the sections below.

Nagashima: Has been working at Mazda since his graduation from the Graduate School of Engineering of Hiroshima University in 1980. Mr. Nagashima has been the Director of the Vehicle Test Research and Development Department since 1980 and the Director of the Mazda Technology Research Institute since 2010. He has undertaken scientific and technological research and development at Mazda for over 30 years. The predecessor of the Mazda Technology Research Institute was the Technology Research Department, established in 1952. The Technology Research Department changed its name to the Technology Research Institute a decade later.

5.2 Hiroshima University and Mazda

Hiroshima University is the core national university in the Chugoku/Shikoku area, and one of its development objectives is to promote and drive the development of the local economy. To achieve this goal, the UIC Center of Hiroshima University plays an active role in promoting collaboration between the university and local enterprises. As the biggest motor enterprise in the Chugoku/Shikoku area, Mazda maintains a close collaborative relationship with Hiroshima University for its technology development. This long-term and stable collaborative relationship provides not only strong technical support for the enterprise's development, but also fertile soil for talent cultivation within the university.

5.2.1 Efforts of a Local University

Hiroshima University is a comprehensive national university, consisting of eleven faculties, eleven graduate schools, and one affiliated institute. This university, which was formally established in 1929, is the leading university in the Chugoku/Shikoku area. Other universities, including Okayama University, Tottori University, and Yamaguchi University, carry out UIC activities by following Hiroshima University's lead. According to a survey carried out by Nikkei Business Publications in 2010, Hiroshima University was ranked first for its brand image in the Chugoku/Shikoku area (Nikkei National Geographic Inc., 2010). Hiroshima University positions itself as an "international and local central comprehensive university" by focusing on the concepts of "education, research, and social contribution." This university carries out activities to play the leading role in a knowledge-based society and knowledge community by adhering to the founding concept of "freedom and peace" and its five principles: "pursuit of peace; creation of new forms of knowledge; nurturing of well-rounded human beings; collaboration with local, regional and international communities; and continuous self-development."

In 1996, the Japanese Ministry of Education launched the First Plan to encourage universities and enterprises to carry out research activities. In the same year, Hiroshima University established its UIC center (now called the Center for Collaborative Research & Community Cooperation), targeting research, education, and collaboration with the society. Its research activities were divided between collaborative research and contract research. In terms of collaborative research, professors at Hiroshima University and researchers from enterprises could share facilities and resources. To increase the efficiency of collaborative research, Hiroshima University established a "comprehensive research collaboration agreement."[1] In terms of contract research, university teachers independently conducted scientific research tasks entrusted to them by enterprises. Within its first 15 years of development, Hiroshima University made prominent achievements in collaborative research. The amount of collaborative research significantly increased and income peaked in 2005 and 2008, respectively (Figure 5.1).

To fulfill its education mandate, the center mainly organized exhibitions, seminars, and public lectures to discuss the potential application value of research seeds, but also supported local enterprises and universities in developing human resources. Meanwhile, the center also provided "entrepreneurship education (起業教育)" courses for undergraduates and postgraduates to cultivate their independent entrepreneurial awareness and ability. Hiroshima University had up to 45 startup ventures until 2010.

In order to facilitate social service, the Center for Collaborative Research & Community Cooperation developed substantial information exchange platforms, including the research idea database "Himawari (ひまわり)" and Liaison Seminars (Hiroshima University, 2013). Enterprises could obtain the latest news related to the university's research activities through the Himawari

[1]"Comprehensive research collaboration agreement" is a systematized UIC mode. When an enterprise has more than one collaborative research with the Center for Collaborative Research & Community Cooperation, such an agreement can be reached to increase collaboration efficiency. Under this agreement, both the enterprise and the Center can assign special people to lead the collaboration team by jointly setting up an operating committee to promote the research through orgnization and joint efforts from points to surface.

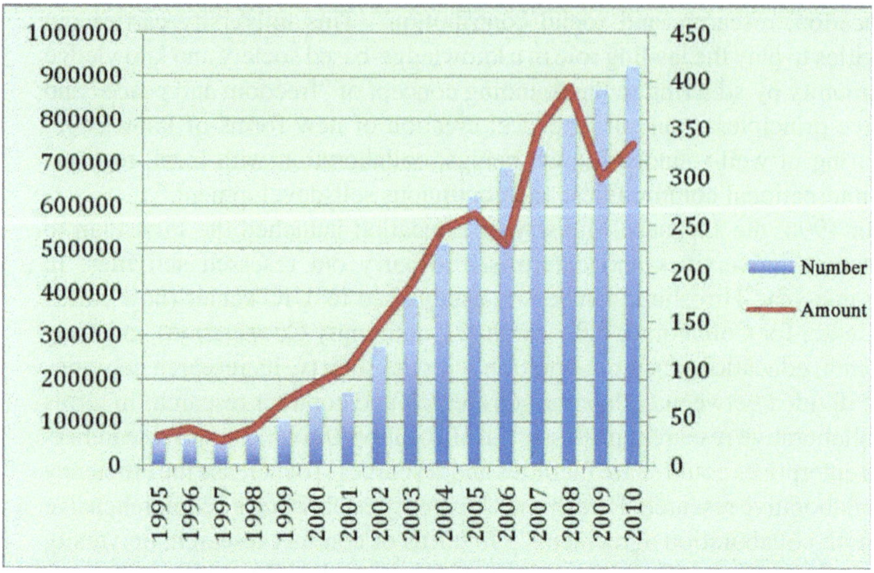

Figure 5.1 The number and income of collaborative research projects (1995–2010).

Source: Hiroshima University (2010). *Annual Report of UIC Center of Hiroshima University in Heisei 22*. Hiroshima: Hiroshima University, p. 21.

site. The Liaison Seminars were mainly convened in Tokyo and Hiroshima to promote the research ideas developed at Hiroshima University. In the seminars, individual discussions could speed up the effective transition of a new technology from a university research achievement to an enterprise application. The university professors with research achievements could participate in discussions as well jointly explore the potential industrial application value of these research ideas. The Center for Collaborative Research & Community Cooperation promoted the implementation of UIC and the development of human resources by cooperating with local governments and industrial promotion organizations. To further community collaboration, the center cooperated with the UIC Conference in the Chugoku Region and other organizations to jointly promote UIC and university-to-university collaboration.

In April 2004, according to the goals of the national university incorporation reforms, Hiroshima University established the Community Cooperation Promotion Organization; since then, the UIC Center has been under its management. In 2010, the Center for Collaborative Research & Community Cooperation was established by integrating the UIC center, regional collaboration center, and community cooperation center of medical care. After more than 10 years of development, the original UIC center has evolved to contain a relatively complete organizational structure. Currently, the Center for Collaborative Research & Community Cooperation consists of five divisions, including the International Industry-Academia Collaboration Division, the Education & Venture Business Creation Division, the Intellectual Property Division, the Community Cooperation Division, and the Biomedical Engineering Research Promotion Division/Hiroshima Branch Office (Figure 5.2).

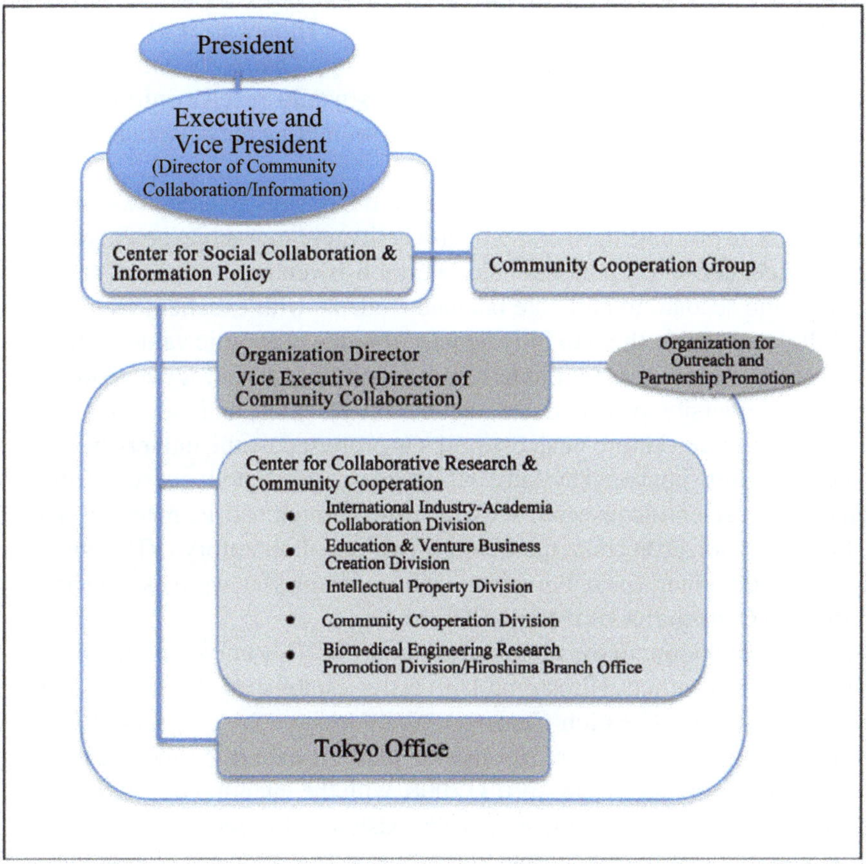

Figure 5.2 UIC organization structure of Hiroshima University since 2010.

Source: Hiroshima University (2010). *Annual Report of UIC Center of Hiroshima University in Heisei 22*. Hiroshima: Hiroshima University, p. 2.

In June 2008, MEXT proposed the "Project of Promoting UIGC Autonomization in Universities," and selected and sponsored sixteen universities for international UIGC promotion. Hiroshima University was one of the sixteen selected, and through this project, its overall promotion of international UIGC began. For this purpose, Hiroshima University established the International Industry–Academia Collaboration Division to develop scientific research potential and strengthen international publicity for its research achievements; in the meantime, overseas offices were set up to promote the globalization of UIC projects. The International Industry–Academia Collaboration Division built the network site for the Center for Collaborative Research & Community Cooperation and the research idea database Himawari. In December 2010, the journal *Hiroshima University Technology Quarterly* commenced publication, ensuring that international partners would receive information related to the latest scientific breakthroughs and promoting the development of international UIC.

The Education & Venture Business Creation Division assumed two main missions: first, to promote the transfer of university research achievements into social practice, and to support and help to establish venture businesses at the university; and second, to cultivate talented students with excellent management ability and enable them to fully benefit from the economic value of their research achievements, such as implementing operating technology education. Hiroshima University provided ten incubation research and development laboratories to serve venture business projects proposed by the university, the venture businesses sponsored by, but established outside the university, and the search for other scientific research activities with entrepreneurial potential. To date, the incubation projects of the Venture-Business Laboratory (VBL) have produced twenty enterprises. Forty-two venture businesses were established up until 2009 (Hiroshima University, 2013).

In 2003, responding to the reorganization of MEXT's university intellectual property head office, Hiroshima University established the Intellectual Property Community Creation Center (知的財産社会創造センター), in accordance with Japan's intellectual property promotion plan. At the same time, it established a "recognized TLO." On April 1, 2008, with the intention of improving intellectual property management and co-existence for regional development, Hiroshima University combined the Intellectual Property Community Creation Center with the TLO, and established the Hiroshima Technology Transfer Center (HTC) (ひろしま技術移転センター), later renamed the Hiroshima Industry Promotion Organization, to perform the function of a TLO.

Although the organization's structure had gone through several changes, it had made gains in patent application and utilization. The number of domestic (Figure 5.3) and international (Figure 5.4) patents applied for increased. The number of registered patents nearly doubled, increasing from 48 in 2003 to 86 in 2007.

The Community Cooperation Division was intended to enhance the mutual development of local communities and the university. It functioned as a liaison between Hiroshima University and local communities. It carried out collaboration not only in research and education, but also in other areas. Its ultimate goal was to "form a new partnership mode between the university and the community," and to boost the growth of the local economy using the knowledge and technologies of professors and students from Hiroshima University. The "research on promotion of local development" began in 2002 and was re-implemented in 2010. Based on the achievements from these research projects, the university provided further support to the research projects that required in-depth expansion.

In the Chugoku/Shikoku area, Hiroshima University first reached an agreement with the local government of Kure City in January 2006. It then reached agreements with, in chronological order, Kitahiroshima, Miyoshi, Higashihiroshima, and Sera. By July 2009, it had established five local

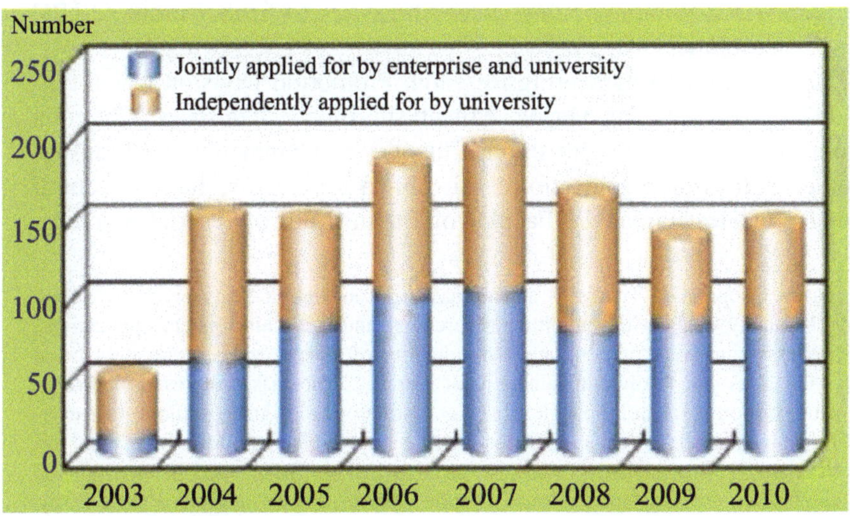

Figure 5.3 Number of domestic patents applied for.

Note: Blue jointly applied for by enterprise and university; *orange* independently applied for by the university.

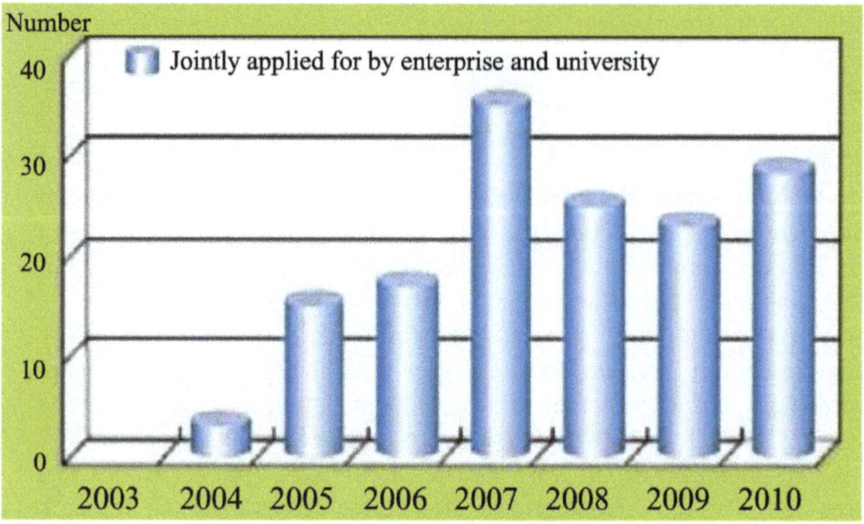

Figure 5.4 Number of PCT[2] patents applied for.

Note: Blue jointly applied for by enterprise and university.

agreements, beginning a new era for the collaboration between universities and local communities. The Community Cooperation Division was primarily responsible for researching successful collaborations between universities and enterprises and summarizing successful cases to further promote UIC in other communities.

The Biomedical Engineering Research Promotion Division was a typical UIGC, consisting of Mazda, Mitsubishi Heavy Industries, Chugoku Economy Federation, the Hiroshima Chamber of Commerce and Industry, and other industrial organizations; Hiroshima University, the Prefectural University of Hiroshima, the Faculty of Engineering of Kinki University,

[2]PCT (Patent Cooperation Treaty) is an international cooperation treaty on patent authorization. It mainly deals with submitting, searching, and examining patent applications, as well as assessing collaborations and the suitablity of technological information being spread. PCT does not "authorize international patents." The task and responsibility of patent authorization is taken by patent offices or institutions (designated offices) that exercise such official powers in countries that seek patent protection. States party to this treaty constitute the International Patent Cooperation Union. These states have priority in patent application, and their internationally applied patents can be authorized more easily. (*Blue* jointly applied for by enterprise and university; *orange* independently applied for by the university) patents applied for.

the Hiroshima Institute of Technology, Hiroshima International University, Hiroshima Kokusai Gakuin University, and other universities; and the Hiroshima government, the Hiroshima Industry Promotion Organization, the Chugoku Industry Creation Center, the Chugoku Center of AIST, and other government institutions.

5.2.2 Independence of Mazda from Outsourcing Service Provider

Mazda is the fourth largest automobile manufacturer in Japan, a famous automobile brand around the world, and the only motor company in the world that researches, develops, and produces rotary engines. It was ranked 255 in the Fortune 500 in 2008. Jujiro Matsuda established Mazda in 1920. The name "Mazda" comes from the name of a god of an ancient religion. Jujiro Matsuda named his company Mazda with the intention of pursuing brightness and goodness. Since the 1990s, Mazda has become one of the most famous Japanese automobile brands in the world, and is ranked behind Toyota and Nissan in Japan.

The Japanese economy developed very rapidly in the 1960s and 1970s. During that time, enterprises had investment capabilities, and many of them began to purchase machinery. Mazda had its own research equipment by then, and no longer relied on universities for experimental equipment. Though Mazda did not rely on universities for experimental equipment, it still required scientific research collaboration with universities for cutting-edge special research projects. Mazda's research ability contributed to its rapid development. After the introduction of the rotary engine by Wankel, Mazda successfully developed the six-air inlet electronic control rotary engine (after technical research and improvement). This engine could control the loading amount with a micro-computer, and automatically adjust the idle device and exhaust gas recirculation device so that it was able to run stably and reduce fuel consumption and exhaust (Figure 5.5).

But Japan experienced the bursting of economic bubbles in the 1990s. In 1995, Mazda suffered huge operational losses for the first time, and was ranked the second-most money-losing enterprise in Japan. The highest annual loss reached JPY 155.2 billion. These losses lasted 6 years. At that time, Mazda's partner, Ford, which held a 20% stake in Mazda shares, stepped in. In 1995, Ford provided research funds for Mazda to research a new gasoline system. Mr. Kawakami said, "It was a outsourcing strategy for Ford. But now, Mazda no longer keeps such a close relationship with Ford." Cutting-edge research was the key to increase the competitiveness of enterprises. Successful

Rotary engine: The rotary engine, which was researched and developed by Mazda, applies hydrogen as fuel and discharges vapor after combustion. It realizes zero pollution. Currently, Mazda is the only company that researches and develops the rotary engine in the world.

Figure 5.5

research and the development of the rotary engine finally enabled Mazda to be independent of any outsourcing service.

In November 2000, Mazda prepared its "New Century Plan," aimed at making Mazda a motor enterprise that could meet different market demands, and ensuring its future sustainable development with a strong product lineup. The plan had four focuses: market, product, finance, and talent cultivation. After implementing the new plan, Mazda's production output decreased by 15%, while its profit increased. In 2002, Mazda's net profit reached USD 65 million, the highest since 1982. In 2008, Mazda had a staff of 18,500 in Japan, manufacturing bases in Hiroshima and Hofu, and a research and development center in Yokohama. It had also built nineteen overseas factories in countries such as the US, Thailand, and such continents as South America, and the Middle East, and a research and development center in the US.

5.3 Internal Combustion Engine Research Collaboration Project

5.3.1 Launch of the Internal Combustion Engine Project

The collaborative relationship between the university laboratory and Mazda was firstly established by Professor Hiroyuki Hiroyasu. However, the initial collaboration was limited to the information exchange and scientific and technological consultations, and featured no substantial cooperation in technical fields. Since 1980s, Hiroshima University began to take on contract research from Mazda. This research was one-dimensional, and the projects were small compared with those of today. The university laboratory earned only about JYP 1 million per project.

Mazda's production system consisted of research, application, and product manufacturing, but before the 1980s, Mazda had no dedicated technology research and development system. Since the 1980s, research, application, and product manufacturing have become independent, and the research department has emerged as the most important. Therefore, the technology research and development department of Mazda represented the company's development in technology and research, which is primarily responsible for research and development related to the Mazda engine. Before 1985, scientific research and development had focused exclusively on the rotary engine. In 1985, Professor Hiroyuki Hiroyasu began a collaborative research and development relationship with Mazda. At that time, the collaboration consisted mainly of Mazda using the university's instruments and equipment to do experiments and learning the university's advanced experimental concepts and methods. But after 1990 Mazda had its own experimental equipment, and its collaboration with the university turned to the application of relatively cutting-edge technology from the university.

Collaborative research began in 1990. At that time, the students from the university laboratory discussed ideas and conducted experimental research jointly with engineers from Mazda. Professor Noda began to participate in the research and development of the jet monitoring system, and then introduced this assessment technology to Mazda. By applying this technology, Mazda began its research and development of a new generation of engines (the CX-5 engine). In 1995, the gasoline system was modified significantly. The modes and subjects of the research collaboration were adjusted accordingly. Noda Laboratory changed jet injection into direct injection (Figure 5.6).

Professor Noda noted, "The real collaborative research began when Mr. Kawakami started studying in Hiroyasu Laboratory." In 1996, Mr. Kawakami, the engineer from Mazda, began his master's degree research with Professor Hiroyasu. Before this, Mazda had generally entrusted a project to the university, and then convened a discussion meeting with the university laboratory regarding the research results. After Mr. Kawakami joined the Hiroyasu Laboratory, the university and enterprise's project teams began holding a discussion meeting each month to exchange information and discuss the next steps of their research plans. After that, the enterprise would send three to four engineers to work with Professor Noda and other researchers almost

Figure 5.6 CX-5: It applies the technology of Skyactive engine. It is the first SUV in volume production under the design motif of "KODO - Soul of Motion" and the symbol of the new era of Mazda brand.

every day in order to keep the information on the laboratory's research results and experimentations up-to-date between Hirosihma University and Mazda. In addition, the subject of the collaborative research between the university and the enterprise gradually became the primary subject of internal research and development at Mazda.

Mr. Kawakami was the principal of the CX-5 engine research and development process. At the beginning, however, he had no relevant technology research background, so he went to Hiroshima University to study this technology. Hiroshima University led Japan, and even the world, in jet research at that time. From 1996 to 1998, Mr. Kawakami studied at Hiroshima University with Professor Hiroyasu, obtaining his master's degree. Since then, the Noda Laboratory and the Mazda Engineering Project Team have enjoyed a closer collaborative relationship. Professor Noda divided the project team into three sub-project teams: a team for research on gasoline spray and its combustion system, a team for research on diesel spray and its combustion system, and a team for research on crosswind spray. The three sub-project teams held meetings with Mazda every month. In addition to these meetings, the engineers from Mazda and students from the laboratory directly communicated with each other on all research data and results. All project experiments were conducted by postgraduates and engineers from Mazda in the laboratory at Hiroshima University. These engineers almost treated the Hiroshima University laboratory as their own. While the project was being implemented, Mazda sent engineers to participate in the experiments conducted at the university every day. Meanwhile, Hiroshima University employed engineers from Mazda to teach classes. Generally, Mazda sent engineers to the university to teach classes for 3 months every year. The funds for this project would be transferred directly from Mazda to Hiroshima University. The Noda Laboratory used the funds to purchase experimental equipment and materials to ensure the smooth progress of its scientific research. The project team convened a reporting meeting every month. Generally, the meeting had three parts: first, reporting on the status of problems left open at the last meeting; second, reporting on the completion of the tasks assigned in the meeting and the experiments' progress; and third, clearly defining the problems to be solved and preparing the research plan for the next month. Since Li was involved in the collaborative project, he was working with the engineers from Mazda as well as with researchers from the university. Li was very nervous at each meeting, as he could never finish his tasks in time and was afraid of being questioned for not finishing: "Generally, I could complete the tasks of the month in time, as long as there were no equipment failures. But each month, the equipment would fail due to various problems, so I could only complete about 50% of my task." During each meeting, Mazda would distribute an agenda for the meeting to each attendee. The project team for the university consisted of Professor Noda and four students from the Diesel Project team, while Mazda's consisted of three engineers, a project principal, a numerical simulation engineer, and a meeting host and recorder. During the first part of the meeting, when the student reported the status of the last meeting's problems, the student would explain the solution using PowerPoint, and the Mazda engineers would ask for clarification when they were confused. In the second part, the student reported the experiments conducted that month. Generally, the Japanese students were responsible for summarizing the experiments conducted by all students in PowerPoint format, which sometimes required tens of pages of slides. This part of the process was the most detailed and time-consuming. During the explanation of each page's contents, Mazda engineers asked questions if they saw problems or thought they could solve problems. Throughout the process, the engineers always had to explain the circumstances of their data. For instance, they would give possible explanations for any strange results. When faced with usual problems, the engineers shared their internal data from related experiments conducted at Mazda in order to explain the original experiment and the differences between it and the students' experiment, and then analyzed the reasons for those differences. Professor Noda sometimes gave his opinion during these discussions. Although this process took a very long time, the students could benefit greatly from it. Li said he learned a lot from these discussions: "I can learn about not only empirical analysis based on practice, but also academic analysis based on theory." After the discussion, the students would summarize the problems that remained and select the next month's research objectives, or "homework," as Li called it. On studying in the Noda Laboratory, Li said, "I have no time to slack off. The enterprise's tight schedule propels my research progress." Currently, regular communication has become the unwritten system; it promotes the growth and development of both sides.

In 2012, the Mazda CX-5, for which Mr. Kawakami mainly developed research, was publicly launched. The CX-5 engine applied the SKYACTIV technology and realized the highest fuel efficiency among vehicles of the same level. The Mazda CX-5 became the first SUV in mass production that employed the SKYACTIV technology.

As discussed above, the production of CX-5 is the successful product of university-industry collaboration, which is a fraction of development of Mazda company. From the viewpoint of development strategy, Nagashima made a long-term plan for UIC in the next 5–15 years. The plan was designed based on the relationships among universities, enterprises, and the government. It listed the scientific subjects that would require collaboration with universities during that period; defined the research direction in the collaboration with universities, enterprises, and the government; and drew an omni-directional and multi-field scientific blueprint involving human engineering, engine manufacturing, etc. The plan may be modified in the future, but the design of the plan was helpful for Mazda to clearly define how a new high technology should be developed in the future, and to enhance its competitiveness in the automobile market.

5.3.2 Collaboration Mechanism

The long-term collaboration between the Noda Laboratory and Mazda is an outstanding case in the development of UIC at Hiroshima University, primarily because of the efforts on both sides to maintain a positive and productive collaboration. With the establishment of the UIC center at Hiroshima University, it served as an intermediary to further promote the overall collaboration between the university and Mazda.

5.3.2.1 Technical integration of the spoke principle

In scientific research and development, the establishment of collaboration with universities was always the main objective of Mr. Nagashima. He compared this mode of collaboration to the "spoke principle":

> The convergent point of spokes at the center of a wheel is the core of the theory. The spokes represent professors from various research fields. To establish cooperation with professors is the key to constructing the whole research network. For instance, research issues on the engine and relevant fields in the collaboration with Professor Noda of Hiroshima University can be solved or jointly researched by famous professors from other universities recommended by Professor Noda. I carry out research on cells. There is a professor at Hiroshima University who is the best cell researcher in Japan. After establishing a cooperation relationship with this professor, I can reach other people who are engaged in research on chemical reactions of cells, because many of this professor's students may be engaged in various research aspects in the cell field.

Mazda's main product is a vehicle, and a finished vehicle is an integration of various parts. In technology research and development, not every part of a vehicle applies the cutting-edge technology that may be required in assembling the vehicle, but some parts require basic research in specialized fields. Mr. Nagashima noted,

> The collaborative research between Mazda and the Japan Aerospace Exploration Agency (JAXA)[3] allows them to learn from each other's technology. Mazda can understand the overall vehicle more clearly, while JAXA can understand the parts of the vehicle more clearly. Collaborative research can integrate the explanation of Mazda's whole vehicle and the explanation of JAXA's vehicle parts. As a result, in collaborative research with JAXA, both sides contributed no capital except technology.

Mazda carried out many similar collaborative research projects, collaborating with different universities on different parts of a vehicle in order to capitalize on the advantages of each university, and then assembling the parts into a finished vehicle. Nagashima believed that "The research of a single university is too narrow, so the scientific research advantages of different universities

[3]The Japan Aerospace Exploration Agency (JAXA): established on October 1, 2003, and enacted by three divisions of Ministry of Education, Culture, Sports, Science and Technology and combined with ISAS, NAL, and NASDA, to form the Japan Aerospace Exploration Agency.

should be combined. Collaborative research promotes the development of Mazda in different fields, and the most famous rotary engine in the world was manufactured in this way."

Mazda has built up a large knowledge network through the efforts of Professor Noda. Mazda would consult Professor Noda when encountering problems outside his area of expertise, and Professor Noda often recommended other famous professors and scholars to help solve these problems. According to Kawakami's recollections,

> By recommendation of Professor Noda, Mazda has established collaboration with Waseda University, Kinki University, and Cosmo Company. At present, Mazda has collaborated with Waseda University for five years, and with Kinki University for three years. Waseda University mainly engages in the chemical reaction of consumption, and Cosmo Company provides various types of gasoline. In these collaborations, Mazda is able to conduct research and further the development of the engine system. This is the collaboration of research teams of the new engine CX-5. By combining all results of the collaborative research, Mazda developed the new CX-5 engine system.

Professor Noda noted that Japanese enterprises had strong scientific research and development capabilities, allowing them to conduct some scientific research and development work similarly to university researchers, leaving them to rely on university professors only in certain fields. The Noda Laboratory was only responsible for the study on the dynamics of engines in its collaboration with Mazda, while Mazda completed most of the engine research and development work independently.

5.3.2.2 Platform constructed by UIC center

From its establishment in 1996, the UIC Center of Hiroshima University actively promoted UIC activities in different subjects. Although the Noda Laboratory had a long history of collaboration with Mazda, Mazda established an overall collaboration agreement with Hiroshima University through the UIC center to better absorb and make use of the university's advanced technology.

Mr. Nagashima noted that the development of electric vehicles would become one of the main research and development goals in future years. In collaborations with universities over many years, good results were sometimes obtained, but generally university professors focused exclusively on their own research, with no intention of transferring their research achievements to become finished products. Nagashima expressed his regret that

> In most cases, the conception of universities conflicts with the design conception of enterprises. If the research achievements of university professors could be integrated with the research of enterprises, the research and development of enterprises would grow more steadily. To change this situation, Mazda takes the initiative to establish cooperation relationships. It organizes discussions on the technology required for future years, such as what kind of cell is required by electric vehicles, through organizations of the UIC center at Hiroshima University, based on the plans of professors in different fields.

The discussion of future research and development schemes with universities has just started. If this new agreement is to have the effect that Mr. Nagashima hopes to see, it will require a joint effort from both sides.

5.4 Win–Win and Conflicts

5.4.1 Realization of Economic and Philosophical Win–Wins

In collaborative research with the Noda Laboratory, Mazda contributed capital to purchase a monitoring camera, jet case, and other expensive experimental equipment. Professor Noda and his students could smoothly finish the experiments and research for the jet system only with the support and guarantee this equipment provided. The large amount of capital required for computer simulation also came from the funds contributed by Mazda for this project.

Professor Noda emphasized that it was more significant that the professors' preexisting scientific research interests increased their desire to collaborate with enterprises than that of the scientific research funds provided by enterprises guaranteed the implementation of scientific research at the university. In the collaborative research for the jet system, professors and their students were able to participate in the monitoring experiments of complete machines in the enterprise laboratory in addition to independent experiments on jet monitoring in the university laboratory. During this process, the professors also gained some practical knowledge they could not learn in universities, laying a foundation for further improvement of academic theory. To share knowledge from collaborative research enabled professors and students to further understand the fit of theory and practice.

Chen said that the professors he knew in the Graduate School of Engineering loved doing scientific research. A teacher at the Graduate School of Engineering impressed him deeply:

> Professor Ishizuka, a professor of the combustion discipline, devoted himself to scientific research, so he remained unmarried. In addition to his own research, he also did experiments in enterprises through collaboration. He attained many research achievements independently. Even collaboration with enterprises was purely out of enthusiasm for scientific research. In my study, I follow the instructions of Professor Noda, my own supervisor, and gain helpful assistance from other professors in the laboratory when encountering problems in relevant fields.

Chen believed that a stable UIC relationship further bolstered the professors' enthusiasm for scientific research:

> The collaborative relationship between Mazda and the Noda Laboratory is relatively stable and close. Every year, Mazda allocates fixed funds for the Noda Laboratory so that Professor Noda doesn't have to seek research funds to support the research in the laboratory from other enterprises. With stable and sufficient funds, Professor Noda is able to focus on combustion and jet experiments, and he need not negotiate with other enterprises due to insufficient funds. To seek research funds, professors have to spend time and energy contacting enterprises and companies as well as establishing interpersonal relationships. With such negotiation work avoided, professors can focus on the scientific research they are interested in. When enterprises are satisfied with the research achievements of professors, they will maintain close collaborative relationships. Once this mutual confidence between enterprises and professors is established, the collaboration will be sustainable, and enterprises will collaborate with the professors when they have research and development projects. As a result, a continuous and stable collaborative relationship is established.

Chen said it was a pity that collaborations between enterprises and universities in China lasted only for 1 or 2 years:

> Universities have to change collaboration partners constantly after completing projects, and it is hard to keep stable and long-term collaborative relationships." He said the collaborative relationship between Shanghai Jiaotong University and AVL[4] was close: "By establishing a joint laboratory just like the joint laboratory for Mazda and Hiroshima University, students and professors in universities and engineers from enterprises are able to participate in research together. Such a collaboration of the joint laboratory is so close that the enterprise will provide scientific research funds every year.

5.4.2 Research Direction Chosen by Students in Practice

At the Noda Laboratory, there were experimental instruments only for jet inspection, and none for complete engine inspection: "Whether the jet research complies with the actual requirements of the engine can be determined only by complete engine inspection. Such a combination of parts and wholes gives students more chances for experimental experience." (Prof. Noda)

Li experienced this practice personally in his experiments. Li mainly conducted jet experiments, and explored the causes of problems in specific experiments

> High injection pressure is required for the engine spray. Generally the pressure should be from 1000 to 2000 barometric pressures, i.e., 100MPa to 200MPa, which must be provided by the system of the fuel feed pump. But after several groups of experiments conducted as per the parameters provided by the Mazda team, the fuel feed pump went wrong for unknown reasons. For safety reasons, the fuel feed pump is equipped with an automatic pressure relief valve. When its safety factors are exceeded, the automatic pressure relief valve begins to release pressure automatically. Since the maximum pressure that the fuel feed pipe can bear is 200MPa, the automatic pressure relief valve begins to release pressure automatically when the pressure exceeds 200MPa. But when the pressure reached 100MPa instead of 200MPa, the machine mistook it for 200MPa

[4]AVL List GmbH, established in 1948, is a famous high-tech group with a good reputation in automobile and engine industries worldwide. At present, it is the largest international high-tech group in the world working on the design and development of internal combustion engines, research and analysis on power assemblies, and the development and manufacture of relevant measurement sytems and equipment.

> by default and began to automatically discharge fuel. After fuel discharge, the pressure in the pipe could not reach the ideal jet pressure, so the expected experimental results could not be obtained. The professor and I were not able to solve this problem, and we contacted the engineers at Mazda immediately. They could tell us some possible reasons, including that it might be due to aging parts, and that some parts would be replaced. Even my professor did not have this kind of experience, so we could gain a lot of practical knowledge from the enterprise.

The collaborative research projects conducted by the Noda Laboratory and Mazda were related to cutting-edge technology: "The doctoral students can learn a lot from the collaborative research, and can easily find a topic for their doctoral thesis that interests them," Prof. Noda explained. With instruction being provided to students by engineers, Professor Noda did not need to go to the laboratory every day, and instead simply met with students regularly. There were no course requirements for doctoral students. They learned through scientific research. The professor instructed students on how to conduct scientific research and on how to discuss the progress and results of their research. Professor Noda believed that, in the process of cultivation of doctoral student through the scientific research instead of classroom instruction, collaborative research would not adversely affect students' learning or professors' instruction, but rather promote students' learning efficiency and alleviate professors' workloads.

For the gasoline project in which Chen participated, Professor Noda and the Mazda engineers first discussed collaboration tasks and then determined the research direction. When Chen's project team had results or achievements to report, the engineer who was responsible for this project would discuss them with the students in the laboratory and stipulate what improvement is still needed. Mazda's research plan was very detailed, specifying which tasks would be completed in which month, and designing all research projects in detail. Chen noted,

> Regular meetings and information exchange effectively promoted study progress. In particular, the exchange, discussion, and preparation of the research scheme in regular meetings provided students with a good learning opportunity. Generally, research by students in universities combines the ideas of one or two people. But in collaborative research with enterprises, universities can share in

> the achievements of the whole research team led by Mazda. The research team of an enterprise is subdivided into small research fields such as basic research on gasoline, experimental research, and simulation research. Moreover, much of Mazda's research is related to cutting-edge science and technology serving future product research and development, so most information is kept confidential. Although the professor does not allow it to be published, I can learn about it.

But in collaborations with the Noda Laboratory, the confidential information could be shared. Chen believed "this was a great opportunity for students to learn about the most cutting-edge findings in their research fields. Although students could not publish the experimental data within these areas of study immediately, it was still beneficial to them and gave them the opportunity to learn a lot. Chen was very excited when, on one occasion, Professor Noda was so impressed by Chen's idea that he advised Mazda to implement it. In his words, he felt "a strong sense of achievement."

Professor Noda wittily summed up the impact of UIC on enterprises, universities, and students with a Japanese proverb: "Three sides, all wins."

For universities, scientific research and education are the priority. Collaborative research provides more subjects for academic theses, and sometimes even opportunities for students to gain patents, increasing the universities' fame. Being well-known attracts more excellent students to universities. Collaborative research also plays a very active role in the instruction and studies of doctoral students. Students can gain experience from conducting research projects with engineers from enterprises, and, importantly, gain practical knowledge this way instead of only theoretical knowledge. Moreover, the subjects of collaborative research are real projects, and students are interested in the real problems that arise in the course of working on such projects. All students in the Noda Laboratory were interested in engines, and that attracted them to be involved in the research projects. Research on cutting-edge science and technology also benefits students greatly.

Universities and enterprises distribute the profits from patents arising from the research achievements of collaborative research according to the proportions specified in the established contracts. Generally, the contracts specify that, if profitable patents or significant inventions are obtained from the collaborative research between a professor and doctoral student and an enterprise, the profits from those patents should be distributed based on the contribution proportions of the two sides. If the university gets 30% of the profits, the enterprise gets 70%. Sometimes, the university will allocate 7–9% of its 30% portion as scientific research funds to reward professors.

5.4.3 Construction of Potential Scientific Research Teams in Enterprise

Many enterprises initially established collaborations in order to obtain excellent future employees. Professor Noda explained that, among the "three sides, all wins," enterprises were the biggest beneficiaries: "Enterprises not only accelerate their scientific research and development in collaboration, but also obtain excellent future employees."

In his collaborative research with universities, Kawakami was particularly impressed with two contributions from the university side:

> One thing to be noted is that Hiroshima University provides new technology for Mazda, especially the introduction of the monitoring system. Moreover, the university sends excellent doctoral students to our company, which is very important to our development. Generally, universities conduct basic research, while enterprises conduct practical research. These two kinds of research cannot exist without the other. In collaboration with Professor Noda, the two sides jointly furthered Mazda's application research. The second is exchange of knowledge. We are good at numerical simulation analysis with a computer calculation system. In the past, this kind of calculation was irrelevant to basic research. But nowadays, the monitoring system of Hiroshima University has been introduced into the numerical simulation analysis of Mazda. We keep focusing on research in the application field so that we have rich new high technology for numerical simulation analysis, but lack understanding of combustion and spray systems and other basic research areas. Collaboration with the Noda Laboratory integrated the numerical simulation analysis that we are good at and research from the university on the combustion system.

In 2012, two doctoral students from Hiroshima University would work at Mazda, a student from the Noda Laboratory and a student studying global marketing. Mr. Kawakami was satisfied with the students from the Noda

Laboratory, and noted that one or two students enter Mazda after graduation from the Noda Laboratory every year.

5.4.4 Reliance of Enterprises on Basic Research from the University

Mr. Nagashima further explained the reasons for collaborating with universities: "Although the technology field of the universities is narrow and the results take a long time to get, universities may spend (three or four) decades on research. Enterprises need to integrate such in-depth research into commodities." The research that Mr. Nagashima was currently conducting was related to the explanation of hydromechanics. The research on hydromechanics was divided into turbulence and rectification. He carried out a collaborative research project on hydromechanics with Hokkaido University, Tohoku University, and the University of Tokyo. The collaborative research project between Mr. Nagashima and the universities had lasted for 8 years. In the collaboration, professors and students sent the data and results to the enterprise after experiments, and a regular exchange meeting was convened every 2 months. The three universities discussed and scheduled the specific meetings with Mazda. The meetings were convened in Hiroshima, Tokyo, and Hokkaido in turn. In the meetings, experimental results and problems were discussed, and the next implementation plan was formulated. Mazda had long-term as well as short-term collaborative research programs with universities. Another of Mr. Nagashima's long-term collaboration projects was "human-mechanical engineering" with Hiroshima University. This project had lasted for about 15 years. Generally, Mazda would conduct a long-term collaboration for basic research, which required a long time to develop, and a short-term (2- to 3-year) collaboration for research related to the product development. Mazda carried out a 1-year collaborative research project on the development of brain waves with Hiroshima City University in order to study a driver's reaction to the approaching obstacles and then design vehicles to coordinate with the driver's reaction, such as automatic stopping when the driver sees any obstruction.

5.4.5 Students' Self-directed Research Plans Impacted by Enterprise Objectives

Li benefited a great deal from participating in collaborative research, but sometimes he could not do what he wanted to do. All experiments were conducted according to parameters provided by Mazda, so Li could not implement his ideas. His experimental design could not be implemented unless he persuaded the professor. Mazda's design scheme was a realizable shortcut that allowed students to approach the leading advancement of technology in their field. Although most of Mazda's research projects were cutting-edge studies that no one had ever attempted before, a high success rate was guaranteed by the years of experience of the engineers and professors directing the projects. Li had to think over the distance between the idea and the practice constantly as his ideal experimental concepts were usually rejected immediately.

5.4.6 Disputes between Sides over Handling of Research Achievements

Mazda and the university still had disputes, despite having collaborated for decades. They might encounter various difficulties in their research, but the ownership of the patents from research achievements was always the biggest conflict between the university and the enterprise. In Nagashima's opinion, "The professors who participate in collaborative research are not interested in the patents, but the university wants the ownership of patents. Distribution of patent ownership and licensing rights always arouses disputes in UIC between universities and enterprises." But for the sake of mutual development, the two sides always tried to find a compromise to solve this problem.

Moreover, as Nagashima noted, "Professors are prone to publish research achievements publicly as soon as possible, while enterprises have to keep much information confidential." Generally, Mazda had priority in handling the research achievements from the collaboration. Therefore, professors were not able to publish their achievements publicly without Mazda's permission. In addition, Mr. Kawakami said,

> The unilateral scientific research experiments of a university professor generally cannot provide the ideal results required by enterprises. This is a common phenomenon in collaboration. Professors are always limited to narrow basic research and hardly care about how to put their findings into practice. To avoid non-ideal results, Mazda sends engineers to participate in the collaborative research with the university to control the research direction of the collaboration.

Mr. Kawakami played a significant role in realizing common collaboration targets.

5.5 Summary

Hiroshima University played an active part in UIC as a core university in the Chugoku/Shikoku area for promoting the development of the local economy. The Center for Collaborative Research & Community Cooperation was established upon the issuance of the *Science and Technology Basic Plan* in 1996, and evolved into a professional UIC promotion team with five divisions, which developed the function of UIC in stimulating the local economy.

Compared with the University of Tokyo, the Center for Collaborative Research & Community Cooperation at Hiroshima University, rather than playing a significant role in large, interdisciplinary, comprehensive projects, conducted more communication work for the transfer of university scientific achievements into external products via the scientific and technological seed platform and effectively promoted the collaboration between university researchers and local enterprises. Another typical difference between Hiroshima University and the other two universities mentioned above (the University of Tokyo and Waseda University) was that the term of office of the coordinator of the Center for Collaborative Research & Community Cooperation was longer than the terms of the directors of the other two centers. Mr. Matsui, with whom the author is acquainted, had worked as a coordinator for 6 years. As a result, Mr. Matsui was very familiar with the professors who were engaged in UIC throughout the university, and helped the author contact many interviewees. Finding potential interviewees was much more difficult at the University of Tokyo and Waseda University, indicating different degrees of intensity in UIC projects at different universities, on one hand, and different functions of UIC promotion organizations, on the other.

6

Conclusion

Collaboration between universities and enterprises features an inter-organization collaboration mode. In essence, however, universities create knowledge, while enterprises apply knowledge. Accordingly, the collaboration between universities and enterprises is an inter-organization collaboration, which is the platform required for product innovation. Most current studies analyze the implementation effects of UIC and its contribution to national innovation based on collaboration results. But in fact, UIC involving inter-organization collaboration is itself more of an innovation than are the collaborative results. Understanding how to begin the collaboration process is crucial to the innovation system. This chapter focuses on summarizing the impact of internal and external environment, effective realization approaches, solutions for problems, and other issues surrounding UIC in research universities based on the cases from the University of Tokyo, Waseda University, and Hiroshima University.

6.1 Conclusion and Discussion

6.1.1 Basic Motivation of UIC: Mutual Demands

Development of the knowledge-based economy expands and deepens collaborations between universities and enterprises. Modern knowledge is produced by various kinds of approaches, and consequently breaking the barriers and limitations of basic research, application research, and development research, and science and technology are becoming increasingly application-oriented, gradually obscuring the boundary between universities and enterprises (Gibbons et al., 1994).

Although the distance between universities and enterprises is being narrowed due to the external policy and institutional environment, effective collaboration is based on the mutual demands of the two sides. The initial motivation for collaboration between universities and enterprises was to seek government subsidies and other resources (Senker, 1998). Federal funds received by professional academic researchers at American universities decreased by 9.4% from 1979 to 1997 (Branscomb, Florida, & Kodama, 2003), and this reduction in funds allocated by the United States government to universities has increased their demands on other funding sources. The situation is similar in Japan. Corporatization of national universities shook the absolute financial security of national universities. As a result, these national universities had to seek opportunities for funding from organizations other than the government. At the same time, a rapid change in markets intensified competition in industries, and it was more advantageous for enterprises to seek research and development innovation in the scientific community than their own independent research and development (Senker, 1998). The US issued the Economic Recovery Tax Act in 1980 to increase tax deductions and exemptions for the industrial research and development, support the research efforts of universities and enterprises, and provide a better institutional environment for UIC (Branscomb, Florida, & Kodama, 2003). According to the analysis of this study, the period of transition Japanese enterprises experienced at the end of the twentieth century was closely related to the difficulties they were facing. When technological innovation became enterprises' only chance for survival, demands for UIC by enterprises gradually became more numerous.

Of course, a mutually beneficial economic arrangement is not the only motivation for UIC. It is evident that both universities and enterprises have other demands. For instance, universities may collaborate with enterprises for the purpose of in-depth exploration of research fields, field testing the correctness of some theory, or establishing cooperation with enterprises to benefit students by increasing work opportunities, obtaining useful knowledge and applying it in teaching, etc. Enterprises also have varying expectations of collaboration with universities: obtaining support from government, conducting explorative research for tracing technology development, or attracting talent at a higher level, among others. The three cases of the study showed the diversity of such demands to varying extents. The study also showed that the ideas and philosophy of universities are another important factor in attracting enterprises.

The author summarized the types of motivation for collaboration between universities and enterprises, according to existing studies (Table 6.1), and discussed the fundamental, driving motivations seen in the case studies.

According to the three cases of the University of Tokyo, Waseda University, and Hiroshima University, the main functions of research universities are teaching and research. Therefore, professors were strongly motivated to collaborate only when the research aligned with basic objectives of the

Table 6.1 Motivation for collaboration between universities and enterprises

	Collaboration Motivation	Literature Sources	Cases Exhibiting This Motivation
University	Guaranteeing research funds	Senker, 1998; Yong S. Lee, 2000	Waseda University, Hiroshima University
	Exploring research fields and promoting teaching	Yong S. Lee, 2000	University of Tokyo, Waseda University, Hiroshima University
	Establishing relationships for students and seeking employment opportunities	Yong S. Lee, 2000	Waseda University, Hiroshima University
	Service to society	—	Reference Cases I and II
Enterprise	Developing new products (product research and development)	Yong S. Lee, 2000	Reference Case I
	Seeking new technology (application research)	Senker, 1998; Yong S. Lee, 2000; Chen Guiyao, 2004	University of Tokyo
	Conducting basic research and sharing risk (basic research)	Yong S. Lee, 2000	Waseda University
	Training staff	Chen Guiyao, 2004; Baba and Goto, 2007	Waseda University, Hiroshima University
	Keeping up-to-date on universities' social philosophies	Gregory Mike, 2005	Waseda University
	Concern for future development of society	—	Waseda University

universities. In terms of using UIC as a source for research funds, the University of Tokyo was not worried about funding, and could get more research funds from national projects, and thus did not have substantial motivation to compete for research funds through UIC. But Waseda University, as a private university, and Hiroshima University, as a local national university, had the same strong motivation to obtain research funds from enterprises and to pursue their own scientific research. In terms of aiding students with future employment opportunities, professors at Waseda University and Hiroshima University always closely linked job opportunities for students to scientific research and teaching, while the University of Tokyo, the best national university, seemed less concerned with job placement. The third function of universities was social service, but in the past, there were hardly enough researchers to carry out collaborations based on social service. Hiroshima University, which aimed to serve the local economy, demonstrated this distinguished function due to its nature as a local university.

The financial support required for this collaboration was crucial to enterprises. Enterprises' prime objective is to seek profits. They would consider carrying out scientific and technological research and development only after attaining stable development and sufficient profits. In the case of the University of Tokyo, ZENSHO had just begun to note the power of science and technology when it reached a certain scale and required further development. With regard to the stages of basic research, application research, and development research, seeking new technologies via application research was a common motivation among enterprises. Demand for product development was usually seen in the development requirements of small- and medium-sized enterprises. Large enterprises viewed the development of basic research on a long-term basis as laying a foundation for their development of application research for cutting-edge technology. Most of the collaborations between Mazda and Hiroshima University were based on application research. The expanding development of ZENSHO was also within the scope of application research. Reference Cases I and II both dealt with research on product development in small- and medium-sized enterprises. As the fourth largest motor manufacturing group in the world, Nissan could support basic research with the strength to develop innovative technology with high market competitiveness. This was the strategy used by large enterprises to succeed in the market, but small- and medium-sized enterprises could not achieve this. Training their staff and attracting talent at a high level were also priorities for enterprises (Baba & Goto, 2007). Students who could work on projects immediately after graduation as experienced staff members were of especially great importance. As a result, cultivation of "experienced students" became one of the teaching responsibilities of universities. This target could be reached through collaborative research with enterprises. Both Nissan and Mazda attracted new high-technology talent in this way.

Therefore, effective collaboration between universities and enterprises was established based on priorities on both sides. These complementary demands were also the motivation for both sides to maintain a long-term collaborative relationship, which was independent from government aid. On the contrary, the government played only an intermediary role as, at most, introducer, guarantor, and witness, as described below. Such complementary demands were based on the actual requirements of both sides and were based on current developments of society and the economy.

6.1.2 Important Guarantee for Stable UIC: Institutional Support

Before the 1990s, most UICs in Japan were established by individuals in a scattered and informal fashion. The wide implementation of UIC at the end of the 1990s was mainly attributed to the *Science and Technology Basic Plan* issued in 1996, which established UIC as a basic state policy. Under the guidance of the policy, Hiroshima University established the UIC center in 1996 to promote the collaboration between the university and local enterprises. The University of Tokyo established DUCR in 2002, and Waseda University established the UIGC promotion center in 2006. The establishment of professional UIC administrative organizations institutionally guaranteed the collaboration between universities and enterprises so that university professors could smoothly carry out the collaborations with enterprises even if they had no experience in conducting such collaborations. The UIC centers were mainly responsible for establishing collaborative relationships, helping university professors handle legal procedures related to agreements for collaborations, and handling administrative matters in UIC to ensure that professors could carry out scientific research collaborations securely and confidently.

The Japanese government issued the Act on the Promotion of Technology Transfer from Universities to Private Business (TLO Act) in 1998, which was the first law in Japan concerning the transfer of research achievements from universities to the industrial circle. According to the law, Japan would establish the TLO, which was an intermediary organization responsible primarily for the acquisition and protection of patents arising from the research achievements of universities, and industrializing these achievements with as little delay as possible. Under the guidance of the TLO Act, the University of Tokyo first established TOUDAI TLO (the former CASTI) in August 1998, making it one of the earliest TLOs to appear after the issuance of the TLO Act. According to the 2012 Annual Report from the UIC Head Office of the University of Tokyo, the number of patents applied for by TOUDAI TLO after corporatization was almost ten times the number before the corporatization of national universities in 2004 (Tokyo University, 2012). TOUDAI TLO became the most successful TLO in Japan. In April 1999, the Waseda TLO was approved by the Ministry of Education and MITI (now METI) and established. It was responsible for the implementation and management of matters related to the intellectual property of the university. From 2005 to 2011, the number of the patents registered (meaning the number of patents applied for and received) in both foreign countries and Japan saw steady yearly gains, and increased over this period by a factor of 11.2 (UIGC Promotion Center, n.d.).

To strengthen the links between TLO and universities, the Japanese government provided funds, equipment, and other support for nationally recognized TLOs; at the same time, TLOs were permitted to use the research equipment in national universities for free. The *Act on Special Measures Concerning Industrial Revitalization*, implemented in 2000, promoted the establishment of TLOs in universities, required the construction of a new system for accelerating UIC in the advanced science and technology fields, and encouraged enterprises to entrust research and development to national and public universities on a long-term basis.

Although the establishment of professional UIC organizations provided some security in the implementation of collaboration, insufficient numbers and deficiency of personnel made the organizations little more than empty decorations. To connect enterprises and universities in terms of technology, AIST implemented the coordinator system for UIC in 2001. As mandated by the system, the Japanese government established the National Center of Advanced Industrial Science and Technology. The coordinator was mainly responsible for matching universities and enterprises who could collaborate on technological projects (from a technology-directed perspective), preparing collaboration schemes, and transferring the intellectual property of AIST to private organizations. TOUDAI TLO, Waseda TLO, and the Center for Collaborative Research & Community Cooperation were all special UIC promotion organizations established according to the national policy. As a coordinator, Mr. Matsui was appointed by the Center for Collaborative Research & Community Cooperation at Hiroshima University. Universities could appoint the coordinator independently, and the project funds were provided by government. The government launched relevant policies and measures to facilitate collaboration for both universities and enterprises.

According to the environment policy in Japan, the government helped universities open various "windows" for external liaisons. Enterprises must provide comprehensive support for collaboration to reduce the inconvenience to UIC imposed by internal research and development departments in keeping their own status and advantages in order to establish stable relationships with universities (Adams, Chiang, & Starkey, 2001). Enterprises usually attempted to establish a long-term partnership through UIC and carry out continuous collaborative research projects. Generally, the longer the collaborative relationship was maintained, the more obligations the two sides would bear on shared resources, making it easier for them to reach agreements on research objectives and decreasing the frequency and intensity of conflicts (Tornatzky & Bauman, 1997). Enterprises could help ensure that both sides reach an agreement on objectives by establishing the UIC-supporting administrative departments and assigning special personnel to communicate and exchange information with universities. In the cases of Waseda University and Hiroshima University, both Nissan and Mazda established independent research and development departments and assigned special personnel for collaboration with universities, improving the prospects for the smooth implementation of projects with regard to the personnel system. In the case of the University of Tokyo, ZENSHO had no development ideas and objectives for its research and development center were newly established. As a result, the principal of the collaboration himself felt confused about how to determine research topics. How could an enterprise carry out good collaborative research with universities if it were not able to determine its own research direction? Enterprises established various external guarantees and promotion systems to facilitate the establishment of collaborative relationships between universities and enterprises. If the two sides lacked coordination and communication systems for interaction, the collaboration would be short and expensive.

6.1.3 Active Impact Factors in UIC: People

UIC organizations could only provide some security in form: people were the impact factor that had decisive effects on UIC. In collaboration with the University of Tokyo, DUCR made substantial preparations that led to the collaboration between the two parties, but its external guarantee function disappeared after the collaboration began, due to the lack of a tracing assessment system. Thus, the success of the next step in collaboration relied upon the activeness of the participants.

The researchers from universities that supported collaboration made significant contributions to the performance of UIC (Tornatzky & Bauman, 1997). In the cases of Waseda University and Hiroshima University, the universities had long-term collaboration experience, so they accepted and got used to the collaboration culture, thus making the collaboration with enterprises easier. In Reference Case I, university professors believed that the scientific research, funding, and service to society were inseparable. Academy and enterprises would share the same research direction and not be confined to the simple product of thesis publication, and would turn into various practical products. This outlook showed a positive attitude toward collaboration. Consequently, the two sides could collaborate with each other pleasantly even on less attractive projects, thereby maintaining a positive collaborative relationship.

Basic research is different in nature from application and development research, so consequently its management measures (Table 6.2) are different. Generally, basic research required more time. Researchers engaged in basic research needed patience to persevere, and requirements for the leading researcher's abilities were high. Application research could be finished in a timely fashion only when teamwork went smoothly. In the case of Waseda University, Mr. Kurihara, as the person with core responsibility for research on fuel cells, finally overcame the technical difficulties through extreme patience. In the case of Hiroshima University, enterprise staff members were often sent to the university to work and conduct experiments with students to guarantee smoothness in the laboratory's work. Development research required higher efficiency so that the team management could meet even stricter requirements. Unified coordination among team members was the key to smooth production. Development research was not popular with university researchers due to its excessive pursuit of efficiency.

6.1.4 Basic Mode of UIC in the Research University: Research

The collaboration examined in the University of Tokyo case did not continue due to the differing motivations and objectives of the two sides. Although common research achievements were obtained, the collaboration was suspended due to a fundamental divergence. In the collaborations at Waseda University and Hiroshima University, the two parties had a long history and built stable collaborative relationships. Moreover, the universities and enterprises kept consistency in motivations and starting points (developing

Table 6.2 Management measures for different collaboration types

Type	Basic Research	Application Research	Development Research
Case	Waseda University	Hiroshima University (University of Tokyo: not applicable)	Reference Case I Reference Case III
Management measures	1. No specific requirements 2. No time limitation 3. No rush in assessment 4. The level of the head of the group is the key 5. In most cases, no fixed requirements on fees 6. Generally no confidentiality	1. With targets and plan 2. With time limitation and flexibility 3. Making on time assessments of the technology 4. Subject selection and organization play important roles 5. Relatively high fees and loose control 6. Some confidentiality	1. With specific targets and strong sense of planning 2. With strict time control 3. Making assessment immediately after completion 4. Coordination and cooperation between the various participating parties are required; roles of organizations and teams are a bigger focus 5. Generally a large amount of fees, input and strict control 6. Strict confidentiality

basic research or application research). According to the 2011 *Investigation on Science and Technology Research*, 73.8% of the research resources in enterprises were spent on development research and only 6.9% on basic research, while 53.2% of research resources in universities were spent on basic research and 9.2% on development research (Figure 6.1). Thus, it was inevitable that the collaborations would evolve, such that the labor was divided along these research specialization lines.

As to the three cases in this paper (Table 6.3), the collaboration projects at the University of Tokyo, Waseda University, and Hiroshima University belonged to different research categories. The development of Nissan's fuel cell was a typical basic research, but such basic research on cutting-edge technology is dominated by enterprises. Generally, it is called "oriented basic research," which features long periods of time, high risk, and good return on investment. Only large enterprises with significant resources can undertake development in this mode, but this type of research is attractive to universities. For this reason, Nissan's engineers were employed by Waseda University as guest professors to impart near-confidential knowledge to the students. Large enterprises desired and relied upon the basic research of universities. They expected to increase their application research and development research via

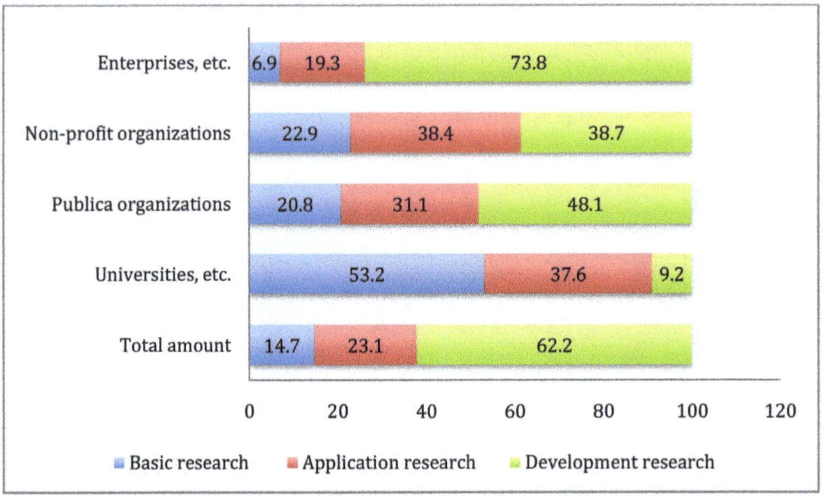

Figure 6.1 Proportion of basic research, application research, and development research of various organizations (2011–2012).

Source: Statistical Bureau of Ministry of Internal Affairs and Communications. "*Investigation on Science and Technology Research, Heisei 23*". http://www.stat.go.jp/data/kagaku/2011/index.htm

Table 6.3 Research types and classification of uic cases

Type	Basic Research	Application Research	Development Research
Definition	Research without specific commercial targets, aimed at innovation and knowledge exploration Basic research with specific targets, conducted using the methods of basic research, is called oriented (targeted) basic research	Research conducted for creating technical foundations for new products, new methods, new technologies, and new materials by applying achievements of basic research and relevant knowledge	Technological research activities aimed at product manufacturing or completing engineering tasks, conducted for creating products, methods, technologies, and materials by applying achievements of basic research, application research, and existing knowledge
Case	Waseda University	Hiroshima University (University of Tokyo: not applicable)	Reference Case I; Reference Case III
Success rate	Generally 50–60% higher possibility of commercialization and enterprization	Generally 50–60% higher possibility of commercialization and enterprization	Generally 90%. The highest possibility of commercialization and enterprization

basic research, from which they had rich experience accumulated for decades in the university. The cases of the University of Tokyo and Hiroshima belonged to the category of application research on new technology development, which is the main type of research in universities. The third kind of research was development research, which was usually carried out by enterprises independently. As shown in Figure 6.1, 73.8% of the research resources of enterprises were spent on development research. This type of research was the least appealing to universities. Although most enterprises carried out development research by themselves, according to Reference Case I, small- and medium-sized enterprises still needed universities to participate in their development research. However, Reference Case III showed that professors were not active enough in such research.

Therefore, research universities should take research as their basic approach to collaboration with enterprises. Research collaboration with complementary advantages can be realized only after universities and enterprises find a common entry point among the three different types of research.

6.1.5 Basis of Collaboration: Differentiation between University and Enterprise

UIC may have positive or negative effects to both sides, and it may also lead to some degree of convergence, but the differentiation between the two sides is the basis of collaboration. For example, in the collaboration between the University of Tokyo and ZENSHO, it was positive for the university professor and his students to directly learn practical knowledge through a visit to ZENSHO, but the collaboration gave the students insufficient time to increase their problem-solving skills due to the requirements of the enterprise on research progress and efficiency. Meanwhile, further exploration of research achievements impeded the timelines of the enterprise in putting developed technology into practice. Therefore, in order to promote technology innovation in enterprises, shall universities allow knowledge creation change to be as market-driven as enterprise's? According to cases at Waseda University and Hiroshima University, universities and enterprises will keep the direction of application research constant. However, what enterprises need most is the abundant basic research accumulated by universities over the course of many years. Basic research is the key to attract large enterprises for UIC. For small- and medium-sized enterprises, the definition of effective research is that universities' application research serves the enterprise's product development, as shown in Reference Cases I and II. These cases also show the difference in social service functions between the leading research universities (University of Tokyo and Waseda University) and the local service university (Hiroshima University). The leading research universities focus on providing support for new high technology and participating in collaboration on cutting-edge technology development. Though the dominant academic subjects offered by in local service universities, they also provide support for technology innovation in large enterprises, and their major function lies in promoting technology innovation in local small- and medium-sized enterprises and the development of the local economy.

Some scholars argued that universities would preferentially invest resources in subjects that create economic benefits faster; that teachers, for the purpose of survival and due to policies promoting UIC after national university corporatization reform, would devote their energy to projects with high levels of efficiency, not projects in which they were really interested; and that teachers' academic freedom would be violated as a result (Tian, 2009). This study showed that professors at the University of Tokyo did not consider economic benefits to be the key factor in collaboration, and that they sought appropriate collaboration partners based on research demands and interest. Hiroshima University, as a local national university, encountered various difficulties in obtaining funds in the competitive system implemented after the reform of national university corporatization, but its professors still considered their research interests and the economic benefits of a proposed collaboration to be equally important factors in their decision to reject or accept the proposal. Compared with the national universities, the private university, Waseda University, was seemingly more deeply affected by economic factors, but its professors' research interests were still considered necessary for the establishment of UIC. In sum, the author did not find that the research interest was totally sacrificed in UIC for the purpose of economic benefit, and thus considers that academic freedom was not violated.

In this regard, due simply to the truth-oriented nature of universities and the benefit-oriented nature of enterprises, the two sides are able to attract and coordinate with each other to realize win–win scenarios in science and benefits and, at the same time, promote social progress and development.

6.1.6 UIC: Begun for Economic Purposes but Concluded with Far-reaching Effects

UIC, as the inevitable route to enable a nation to develop a knowledge-based economy, should generally have two resulting products, knowledge and economic benefit. However, UIC is always dominated by enterprises, so the motivation of economic benefit is more prominent. For knowledge innovation to occur, universities must take the initiative to find suitable cooperative partners for the renewal and creation of knowledge. As a result, institutions involved in UIC focused exclusively on its economic benefits for a long time, ignoring the creation of knowledge.

When universities and enterprises are united in the pursuit of economic benefits, they sometimes disagree about the allocation of achievements. In all UIC, the two parties are faced with the question of how to share research results. To avoid later conflicts, a statement is generally put into the collaboration agreement before the cooperation begins, further increasing the barriers to begin a collaboration. This statement clearly outlines both parties' goals, making it easier to reach a consensus and start cooperation. In the cases of Waseda University and Hiroshima University, basic consensus on the sharing mode of achievements had been reached over years of cooperation, so there were no arguments on this subject. But in the case of the University of Tokyo, since there were a few cases of cooperation in which the researchers and enterprises involved were new to these collaborations, it was not surprising that the two parties failed to reach a consensus. For research results, both parties can use the method of classified allocation: among theses, patents, and principle models, which are the forms that application research results typically take, theses are the results in which university researchers are chiefly interested, while patents are the results that interest enterprises. Reasonable allocation of research results can easily ensure continuing cooperation. When the two parties lack a shared interest in the pursuit of economic benefit, the cooperation between them tends to be difficult to sustain. Universities usually give up on the cooperation which is only interested in enterprises' economic benefit. Eventually, finding an appropriate balance between economic benefits and universities' expectations becomes an issue that enterprises must consider, or risk losing the collaboration project altogether.

Since the beginning of the twenty-first century, in the increasingly competitive era of economic globalization, enterprises have paid more and more attention to ideas and philosophy of universities research (Gregory, 2005). In addition to economic benefit, the ultimate goals of Nissan's collaboration with the Waseda University included a concern about the university's research philosophy. Nissan also hoped that the university could guide the development of the enterprise, which reflects the company's awareness of its social responsibilities. Universities' philosophies and ideas can help enterprises become attuned to their sense of responsibility and mission. Mazda also strives to develop new products for the construction of a green and environmentally friendly society. At the moment enterprises are taking on corporate–social responsibilities, but the author cannot help asking a question: what are universities' social responsibilities?

In UIC, enterprises aim to create more economic value to drive the economic development of society. Should universities find their own positions in societal development? Should universities, especially today's research universities, devote more effort to knowledge creation for the benefit of society? One must let the enterprises' economic effectiveness and the benefits of universities' knowledge jointly drive the development and advancement of society.

6.2 Construction of Effective Cooperative Mechanisms

This study was designed to identify the typical features of effective UIC. Based on cases from the University of Tokyo, Waseda University, and Hiroshima University, the author studied a number of typical UIC examples among research universities in Japan. When the partners take some but not all necessary steps to ensure effective collaboration, the models of collaboration are mixed, with some successes and some failures. In these mixed models of collaboration (between the new and old goals of development, conversation, and conflict and the new and old sources of funding), the matter of understanding and balance is especially important. Besides their complexity, the description of each case study has explored the unique collaboration format of each university. The author's purpose remains to be close to reality at the risk of being trivial. Now the author will endeavor to fully explain how research universities establish and sustain effective cooperation.

Universities and enterprises, as two separate entities, need to work together to achieve effective collaboration. Burton Clark, an American scholar, and the first to present the triangle of coordination in higher education systems, explains the interactions among universities, governments, and markets at a macro level. Henry Etzkowitz' Triple Helix theory also indicates that economic development serves as the link connecting government, market, and university. The strengths of the three parties should be combined for mutual impact and common growth, through mechanisms like organizational and structural arrangement and institutional design, so that they can achieve the goals of resource sharing and effective communication and maximize their individual benefit.

This study suggests that, at a macro level, a common growth model is formed among universities, government, and industries based on interaction, but at a micro level of practice, government plays a background role by facilitating communication and coordination. The course of UIC can be likened to a process in which two groups jointly plant a fruit tree for which organic soil is the basic condition for growth and maturity, while effective division of labor between the two parties, and the intelligence and perseverance of key individuals, are critical elements required for it to thrive (Figure 6.2). Three elements of organic cooperation mentioned below can help to explain the cases discussed in this study.

6.2.1 Source of Organic Soil

The first element is the multiplex platform of supporting systems, including (i) policies supporting UIC issued by the government, (ii) establishment of administrative agencies specially providing support, and (iii) accumulation of professional facilitators for UIC.

To promote collaboration between universities and enterprises, from the formulation of a national strategy and the *Science and Technology Basic Plan* to the *Act on Special Measures Concerning Industrial Revitalization*, the *Law for Special Regulations Concerning Educational Public Service Personnel*, and the *Act on Promotion of Technology Transfer from Universities to Private*

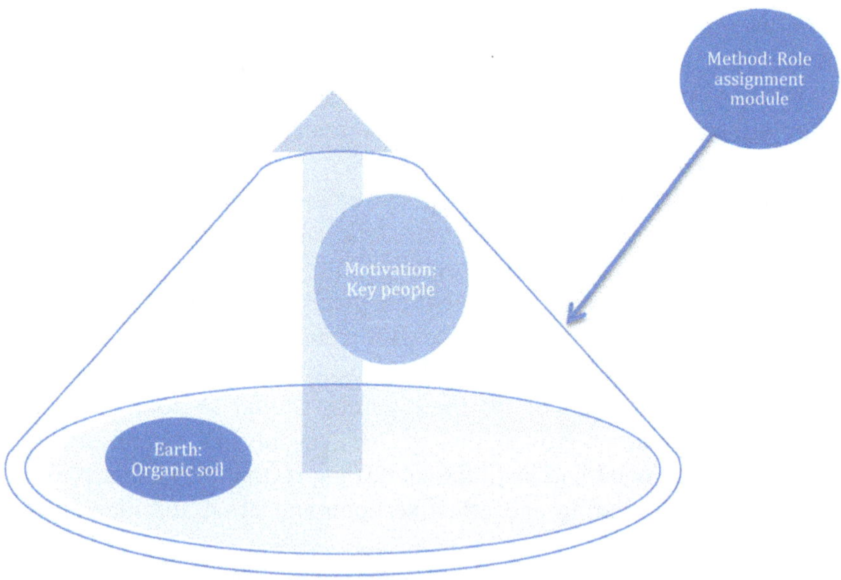

Figure 6.2 Mode for organic (effective) collaboration of the three elements.

Business, the government put forward policy-supporting measures concerning not only industries but also teachers and universities. The policy environment for UIC was, therefore, basically formed. Driven by these policies, local governments and universities began setting up related promotion agencies. Meanwhile, the central government proposed a system of dispatching coordinators to guarantee the equipment of professional personnel in those promotion agencies. The specialization of staff in promotion agencies of UIC directly determines the depth and breadth of universities' UIC.

The mission of the Division of University Corporate Relations (DUCR) of the University of Tokyo is to carry out the training system of the Technology Liaison Fellow (TLF): to train UIC coordinators all over the country to build UIC relationships through on-the-job training. The great achievements made by the University of Tokyo in UIC resulted in part from its large number of professional personnel, all working in the eight-floor UIC building. The length of employees' terms of service is typically 3–5 years, and all of the employees work full-time. The UIC Center of Hiroshima University occupies only half a floor, with another, separate building serving as the Venture Business Laboratory. Although its facilities are relatively complete, the staffs are mainly part-time employees with high-staff turnover. The Science and Technology Incubation Center of Waseda University occupies only a corner of one floor, and about two-thirds of its staff are part-time employees.

The construction of professional teams is the organic soil for the collaboration between universities and enterprises, and a lack of nutrients in the soil nutrients noticeably affects the thriving of the fruit tree.

6.2.2 Key People

The second element is the key people of cooperation teams. Given the same institutional conditions, why is some collaboration carried out effectively and enduringly, while some collaborations end prematurely and even acrimoniously? As the comments of some scholars, on Arima Akito, one reason for the delay of Japan's corporatization reform is due to the lack of talents like Arima Akito, with experience in multiple roles and the resulting ability to understand pursuing multi-party benefits (Hatakenaka, 2004). Over the course of collaborations between universities and enterprises, only people with this variety of experience can really play the role of key persons in cooperation.

Sometimes when a person's position changes, the whole course of a collaboration is affected. Kurihara, an engineer at Nissan, was responsible for a project in the US that halted due to personnel alterations: "At that time,

I started a collaboration with Stanford University, but then I was transferred back to the head office of Nissan to succeed my director and continue my previous work. Finally, this collaboration had to be stopped." A Collaboration means establishing relationship, and the connections among some key persons become more important than the relationships among other people involved. A change in the collaborators is a common reason for the end of a collaboration.

It is also important to withstand the test of time. Enterprises usually focus on market orientation, technology research, and development that can be completed quickly, but innovative scientific research and development requires a certain growth cycle. Perseverance and patience are the keys to ultimate success in this area. Kurihara repeatedly stressed that it is critical to value intermissions in scientific research. When others forgot fuel cells, Nissan put them on its research agenda again and went forward with the research and development of fuel cells. The Nissan Leaf was the result of this persistence and effort.

Kawakami of Hiroshima University, and Kurihara of Waseda University, had worked for both universities and enterprises. Based on their understanding of these two different organizations, they could find the right balance in a collaboration and then effectively push the UIC process forward, ensuring that the collaboration would continue. Over the course of their development and growth, enterprises would often periodically select suitable internal candidates and send them to universities to study for professional degrees. Enterprises hoped to enhance mutual understanding and construct effective collaboration mechanisms in this way. However, it was not easy for enterprises to select suitable internal candidates. Universities' doctoral graduates, in contrast, can very naturally serve as coordinators of the collaborations between enterprises and university laboratories when they are hired by enterprises. In the current collaboration between Mazda and Hiroshima University, many projects are directly managed by graduates of Hiroshima University. The way that graduates assume responsibility for communication between the enterprise, and their alma mater is based on their understanding of undertaking multiple roles (in enterprises and universities), and this allows them to effectively promote UIC. In addition to effective communication based on dual identities, key people are also required to have application-oriented vision and future-based philosophical thinking. The direction of enterprises' development and the philosophical value of universities can, therefore, be in harmony with each other, and real technological collaboration and innovation can be achieved.

6.2.3 Mode of Task Division Depends on Research

The third element is designed as a tool for summarizing the collaboration methods for research between universities and enterprises. In the field of human service, specialized services are constantly emerging. These services, the so-called classified projects, are aimed at specific people, specific demands, or specific services. This sectorial structure not only makes cooperation necessary, but also sets up numerous barriers. Resource Dependence Theory indicates that although sectors generally value decision-making rights and strive to avoid cooperation, a sector will have the will to cooperate when its resource base relies on another sector. Based solely on mutual demands and development, universities and enterprises seek opportunities to collaborate with each other.

Universities, enterprises, and governments perform interactive development at a macro level. But as we focus on the micro practices of UIC, the three subjects of the triangle mode do not appear and exist simultaneously. The *Y*-axis represents Enterprises (green), and the *X*-axis represents Universities (black). The *dotted portion* of the first quadrant represents Government (Figure 6.3). From the UIC practices of three research universities, i.e., the University of Tokyo, Waseda University, and Hiroshima University, it was discovered that the government played an invisible role and usually acted as the middleman. Universities and enterprises severally undertake basic research, application research, and development research, or jointly undertake collaborative application research and development research.

The fact that universities and enterprises sequentially undertake basic research, application research, and development research means that enterprises share a rational division of research tasks with universities, according to their own development requirements. In the mode of division, Yb, Ya, and Yp, in the second quadrant, represent an enterprise's independent research, while Xb, Xa, and Xp, in the fourth quadrant, represent a university's independent research. The nine blocks in the first quadrant represent various combinations of collaborative research between enterprise and university.

As enterprises' research in the second quadrant combines with universities' research in the fourth quadrant, a typical form of contract research is established between the two parties. For example, large enterprises like Nissan generally undertake independent application research (Ya) and development research (Yp) on some projects with sole reliance on universities' basic research (Xb). Thus, an Xb/Ya mode or an Xb/Yp mode is formed. Contract research between universities and enterprises is generally formed in the Xb/Ya mode and the Xa/Yp mode.

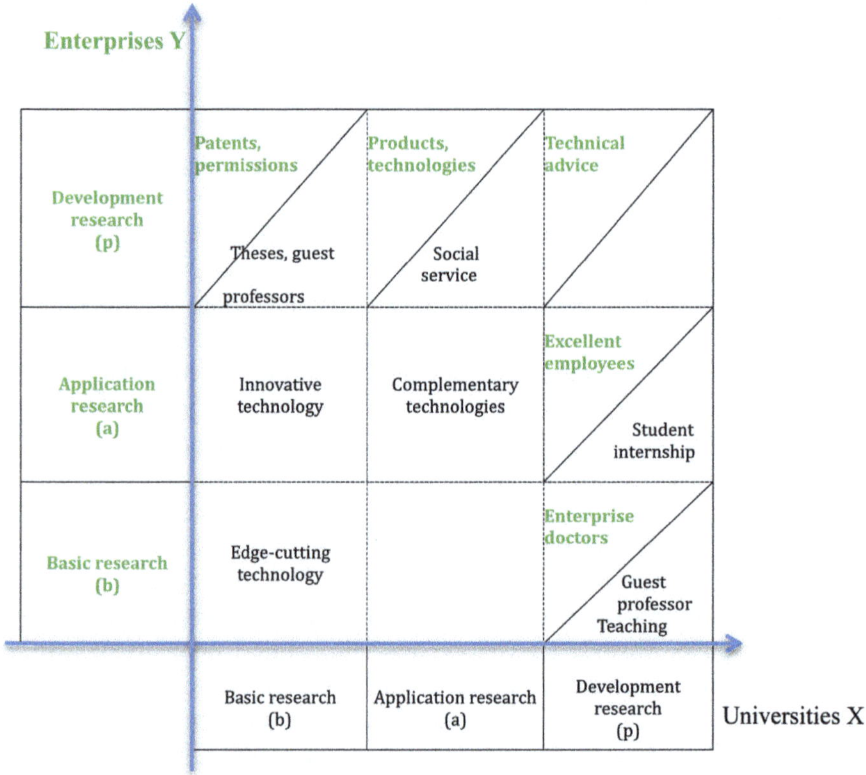

Figure 6.3 Method: Role assignment in UIC.

Source: Internal information from the UIGC Promotion Center at Waseda University.

To clearly explain the nine blocks in the first quadrant, the definitions of basic research, application research, and development research should first be clarified here. On this basis, the forms and achievements of collaborative research between university and enterprise will be presented.

Basic research refers to research focusing on innovation and knowledge exploration not for any specific commercial purpose. However, basic research with specific targets is commonly conducted in enterprises for economic gain. Nissan's research on fuel cells is an oriented basic research undertaking within the enterprise. Application research is based on the use of outputs and related knowledge from basic research, and is performed to study the technological foundations for the production of new products, new methods, new technologies, and new materials. Development research refers to technical research activities through which new products, new methods, new technologies, and new materials are created on the basis of the achievements of basic research, application research, and existing knowledge, and which are performed for the production of products and the completion of engineering tasks. Generally, enterprises consider new products and new technologies to be the goals of development research. Researchers holding new technologies are also carriers of new knowledge in an enterprise. Development research is most clearly seen in the following two forms in universities: the same research on new products and new technologies as enterprises, typically performed by universities and enterprises together, since few universities have their own independent development research; and knowledge products, including textual products like theses and writings and knowledge products as carriers, such as graduates and enterprise doctoral candidates, as well as teaching and training activities in universities. Universities' development research, discussed below, typically takes the latter form, i.e, the form of knowledge products.

In the collaborative research of the first quadrant, enterprises' basic research (Yb), application research (Ya), and development research (Yp) combine with universities' basic research (Xb), application research (Xa), and development research (Xp) and form nine different combination modes of collaborative research. In accordance with the trend of Japan's UIC being dominated by enterprises, collaborative research between enterprises and universities will be discussed on the basis of the Y-axis in the following paragraphs.

First is the combination of enterprises' basic research and universities' research. (i) Intra-enterprise research institutions rarely conduct time-consuming and high-risk basic research, and only large, powerful enterprises formulate a development strategy that spans 10 or even 20 years and conduct basic research. Therefore, this basic research is usually combined with universities' basic research to form the XbYb mode for cutting-edge technology research, such as the research on fuel cells jointly performed by Nissan and Waseda University. (ii) Since the combination case of enterprises' basic research and universities' application research has not appeared in this study, the XaYb mode will not be discussed here. (iii) The combination of enterprises' basic research and universities' development research (the XpYb mode) mainly appears as a carrier of knowledge products. In such a mode, senior engineers working on enterprises' basic research are usually hired by universities as guest professors to teach courses for postgraduates and to promote talent cultivation at universities; enterprise personnel eventually return to enterprises carrying knowledge products after advanced study or after completing a master's or doctoral degree at a university and promoting the progress of enterprise within basic research.

Second is the combination of enterprises' application research and universities' research. As the basis of development research, application research is the research priority for large enterprises. Therefore, the combination of enterprises' own application research and universities' research is the most common form of UIC. (i) By virtue of universities' basic research, enterprises promote their own application research (the XbYa mode) and typically focus on the development of innovative technology. The cooperation between ZENSHO and the University of Tokyo for the development of innovative technology was aimed at realizing a successful enterprise transformation. In addition, several cooperation projects between Nissan and Waseda University were also performed in such mode, of which a typical example is the research on the fuel cell engine. (ii) The combination of enterprises' application research and a university's or scientific research institution's application research (the XaYa mode) is largely based on the demand for complementary technologies. The purpose of the collaboration between Mazda and the Japan Aerospace Exploration Agency (JAXA) is to learn about each other's best technologies. Furthermore, in the automobile industry, concerned enterprises need to learn from researchers in many different fields for the advancement of their technologies so as to improve the performance of the complete vehicle. Therefore, the combination of the universities' application research in related fields is the most common UIC pattern for enterprises. (iii) The combination of enterprises' application research and universities' development research (the XpYa mode) mainly takes the form of students interning in enterprises or enterprises employing outstanding university graduates. Generally, doctoral graduates will be hired directly into enterprises' departments of application research and development to work on scientific research. In the collaborative research of enterprises and universities, students are usually allowed to enter enterprises' departments of application research. Enterprises' application research is confidential to some extent, but is not as strictly controlled as development research. Consequently, students engaging in application research abide by the enterprise's confidentiality requirements. Although relevant information is unavailable for the public students are still able to acquire a lot of advanced knowledge of science and technology, as well as practical experience.

Last is the combination of enterprises' development research and universities' research. For large enterprises, development research is usually considered a trade secret, and is not conducted in collaboration with universities. However, large enterprises still occasionally consult universities for technical advice. Meanwhile, small- and medium-sized enterprises take the combination of their development research and universities' application research as their research emphasis. (i) There are rare cases of the combination of enterprises' development research and universities' basic research (the XbYp mode), but for the achievements made through such modes, enterprises have their own patent rights or permissions. University professors always present such achievements by means of theses or articles, which are presentations of enterprises' universities' joint research (the XbYa mode). Engineers in enterprises can be hired by universities and engage in universities' basic research as guest professors. (ii) The combination of enterprises' development research and universities' application research (the XaYp mode) is generally the first choice for small- and medium-sized enterprises. With the aid of universities' research strength, they often develop new products and profit more from their development. As a result, this mode is more market-oriented. Reference Cases I and II are typical XaYp modes. Universities show their desire to serve society through such combinations. For example, Professor Watanabe in Reference Case I is very hopeful that his research will serve the public. (iii) In combining enterprises' and universities' development research (the XpYp mode), both sides put little effort into new products and new technologies. In most cases, enterprises consult universities for technical advice and gain intellectual support from university professors.

As to the form of achievements, the mode of division is not limited to a few of these blocks but is applicable to most of them. For example, patent rights and permissions are applicable in the XaYa mode. Examples of the core features of each block have been noted. The blocks also show there are many overlapping characteristics in the initial stages of collaboration.

The particular forms highlighted in the above discussion are based primarily on the findings of this study, but it is believed that a variety of forms exist in the actual practice of UIC. As long as collaborative research is carried out on the basis of research efforts, regardless of changes to the forms, it can get a foothold in the above mode of division. Since few cases are discussed in this study, the author cannot say with confidence that the mode of division has full explanatory power for all UIC activities. However, the mode can at least provide specific guidelines for microcosmic practices to enable participants to be free from hesitation and confusion at the outset of collaboration.

6.3 Drawbacks of This Study

Due to limited time, only three cases are discussed in detail in this study. Further discussion and analysis of more case examples is required to prove the universal legitimacy of the effective cooperative mechanisms identified. Moreover, this study was originally designed to demonstrate the organic interactions among universities, enterprises, and governments. However, the invisible roles of the government found in the study piqued the author's research interest. At the end of her research period, Mr. Hosaka established contact with government officials from Japan Science and Technology Agency (JST), but the author failed to perform further research into the reason for the invisible roles of government due to limited time. Certainly, the reason will provide a topic for further investigation for researchers in the future.

Annexure

Annexure 1

Interview Outline

1. Greet and give consent/information form
2. Offer opportunity to ask questions
3. Begin interview

   Lettered questions indicate main questions, while the sub-questions are prompts only to be used if needed. This is only a guide; any question or prompts that seem relevant at the time may also be asked.

   A) What motivates the researchers/company/government to participate in the university-industry collaboration?

      i. What are some potential elements of the project that may act as conditions for your involvement?
      ii. What are the differences between the individual and organizational levels?
      iii. Can you elaborate further?
      iv. Discuss the differences between the motivating forces of your participation from the academic (knowledge/technology/institution innovation), economic, and social (reputation/public value) perspectives.

   B) Could you describe the most impressive case in all the university-industry collaborations in which you have been involved?

      i. Avoiding specifics, can you describe some elements of the case that made it significant to you?
      ii. How did you feel about the success of the collaboration?
      iii. Can you describe how you furthered the establishment of that effective collaboration?
      iv. Describe how you feel when you find what you think is an effective and successful collaboration.

C) Could you describe the least successful case in all the university-industry collaborations in which you have been involved?
   i. What did you do when you recognized that the collaboration was unsuccessful?
   ii. Can you elaborate further?
   iii. How did you feel about that unsuccessful collaboration?
   iv. Is it possible to eliminate or prevent the specific issue(s) that led to this case being unsuccessful?

D) Is there a case that you remember particularly strongly from this collaboration project?
   i. Please describe a significant case in which you experienced among many collaboration projects.
   ii. How did you feel beforehand?
   iii. How did you feel about that?
   iv. How did you feel afterwards?

E) Could you describe a time in which you were working on a particularly difficult or harrowing case?
   i. How did you feel beforehand?
   ii. Did you make any changes to how you approached the project based on your feelings?
   iii. How did you feel about that?
   iv. How did you feel afterwards?

F) Could you describe any pressures that you might have experienced during a university-industry collaboration?

G) Can you describe the checks and balances an organization may use to ensure a high-quality collaboration? What ensures that conflicts do not occur?

H) Could you describe the benefits you received from this project collaboration?

4. Give debriefing form.
5. Give another opportunity to ask questions.

Annexure 2

Reference Case I: Hiroshima University and Revival of Pearl Star Ltd. ("Pearl" for Short)

President Miura (三浦) is 60 years old, with a medium stature and kind face. He began running his father's group of hosiery manufacturing businesses in 1978. For someone like me, one who always maintains frequent contact with large enterprises, visiting such a small company provides a very different experience. Pearl is, in fact, even smaller than I had anticipated, and has only thirty-six employees. President Miura's office is only about 10 m^2 in size, and the showroom, which doubles as a meeting room, is approximately 20 m^2 in size. At the sight of the scale, I cannot help but sense a degree of difficulty and hardship for the development of such a small enterprise.

In spite of Pearl's smaller size, it has run smoothly in the past. However, as the Japanese hosiery manufacturing industry has increasingly shifted overseas, it has become harder for Pearl to develop in Japan. It experienced two significant financial deficits in 1999 and 2002, with the most serious financial crisis since its establishment in 2007 (Figure A1). In 2006, Pearl was already operating with difficulty, which made President Miura to apply a bank loan.

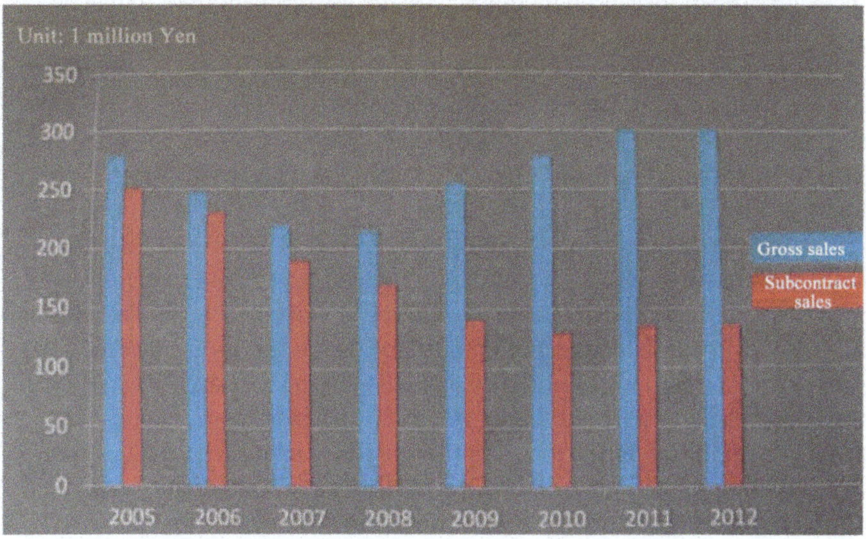

Figure A1 2005–2012 Annual sales of Pearl (1 million yen).

Source: The internal information of Pearl. *Blue* gross sales; *red* subcontract sales.

Prior to that, Mr. Miura had not known where to find podiatry experts and had been inquiring in various places. By chance, he heard about Professor Watanabe of Hiroshima University while at the bank. The men happened to be introduced through a member of the bank staff, Fujii, who had been the bank's representative when Professor Watanabe purchased his house. Mr. Miura was very grateful for that accidental chance, because most people, including those running medium-sized and small enterprises, did not know about UIC at that time. According to Mr. Miura's impressions, UIC had begun during the period of the Koizumi Cabinet (2001–2006), and had a history of only 10 years; it had not been allowed when he studied for a master's degree 40 years ago. Therefore, Mr. Miura was very grateful for the cooperation opportunities available for helping medium-sized and small enterprises out of the developmental stalemates.

The history of the cooperation between Pearl and Hiroshima University was not long. It started with Mr. Miura's meeting with Professor Watanabe in 2006, and collaborative research formally began in 2007. Prior to that, Pearl had no experience in cooperation with other universities, although it did have established contacts within the government. In 2004, Pearl applied for the METI and SEM Support for subsidies in order to purchase machines and develop products. Mr. Miura applied for these subsidies himself, which was the beginning of Pearl's cooperation with the government prior to its UIC with Hiroshima University. In 2006, Pearl encountered financial problems, and the development of new products was required to promote the company's growth. Mr. Miura wanted to research and develop the "fall prevention socks," but without professional guidance, he failed to realize the full potential of his hosiery manufacturing technology. This changed when he met Professor Watanabe and benefitted from the aid of theory. In 2007, with the help of the Hiroshima Industrial Promotion Organization, President Miura and Professor Watanabe applied for the JST project of the Japanese Ministry of Education and obtained financial support in the form of millions of yen (specific amount confidential). Funds provided by the government and donations from companies in the Chugoku area were under unified management of the Industrial Promotion Organization, while other funds came out of Hiroshima's tax revenues. The Industrial Promotion Organization was primarily responsible for the allocation and management of project funds. Pearl, Hiroshima University, and the Hiroshima Industrial Promotion Organization worked together and achieved the idea of "fall prevention socks."

Operators of small and medium enterprises have to consider numerous factors, many of which cannot be fully considered and it is not easy to collaborate with universities. President Miura was glad that his company employed relevant technicians, as the cooperation might have reached a sticking point otherwise. Professors, as researchers, often use many technical terms that intra-enterprise technicians needed to understand in order to advance product development, meaning that enterprises are required to have their own talents with a considerable professional foundation. Pearl succeeded because it had professional talents experienced in hosiery manufacturing. However, in most small and medium enterprises, many people do not know where to find suitable professors, and they likely would not know how to communicate with professors upon meeting them. University professors and small- and medium-sized enterprises come from two different worlds, and combining these worlds effectively is the key. Mr. Miura's experience tells us that both parties can work in conjunction to develop good products, provided that they communicate smoothly. In the cooperation, Professor Watanabe was responsible for foot testing the socks and identifying problems via the results, at that point President Miura called relevant technicians to work out solutions by examining the problem data. Communication between both parties was repeated in the same way. Sometimes, enterprise personnel would perform foot testing and discuss improvement programs with Professor Watanabe. During the research, Professor Watanabe would go through the test cycle repeatedly and then monitor the try-on effect together with Pearl's engineers. The monitoring process used high-speed photography equipment, which could capture every single move made, while the test subject was walking in order to check the friction and balance of the socks. The subject walked forward, upstairs, and downstairs without stopping, and walked on floors made of different materials to check the actual effect of the socks. After 8 months of research, the "fall prevention socks" won the excellence award at the Fukushi Machine Competition in October 2007. Professor Watanabe and President Miura were happy to have solved problems for the elderly who had difficulty in walking and a need for assistance.

It was initially believed that UIC would succeed as a whole, but there were many cases of failure. As Mr. Miura believes, the market is broad only when enterprises' technologies combine with universities' theories. Professors know about foot motor function, but they do not necessarily know how to make socks that sell well, and universities' achievements in scientific research cannot make profits without enterprise. To realize an effective collaboration, President

Miura particularly stressed the importance of clear targets. The acquaintanceship and collaboration between Mr. Miura and Professor Watanabe was accidental. In Mr. Miura's view, he was lucky. If he had been introduced to an unsuitable professor, he would not only have been embarrassed to refuse his help, but also would not have been able to obtain the ideal results. Currently, there are more and more approaches being introduced to provide a chance for enterprises to learn the university's technologies. These approaches build a "rainbow bridge" for small- and medium-sized enterprises. Since 2004, the Hiroshima Industrial Promotion Organization, the Chugoku Economical Industry Bureau, and the SME Industry Organization have held more and more technical seminars, and as the number of these seminars increased, more and more enterprises began to participate. Mr. Miura believes that the seminar is the best platform for the discussion that allows enterprises to communicate easily with professors and learn about those that might be suitable for specific projects.

Currently, the overall situation in Hiroshima is that various support systems are provided for the development of small- and medium-sized enterprises, but there are still some elements that need improvement. Matsui, a teacher at the UIC Center of Hiroshima University, is specifically responsible for proposing collaboration between universities and enterprises, but there are some limitations present in the form of variance in professional fields. Matsui majored in engineering, which is different from Professor Watanabe's professional field, so it may be difficult for Matsui to form sufficient contact with professors in other areas of research. Moreover, the geographic dispersion of the Hiroshima University campus also puzzles small- and medium-sized enterprises; part of Hiroshima University is located in Hiroshima, while another part is located in Higashi–Hiroshima. Collaborations with Higashi–Hiroshima are managed by Matsui, while those with Hiroshima are managed by others. Many enterprises are puzzled about which one to consult. Mr. Miura suggests that business could all be handled at the same place for the convenience of information exchange and to prevent the problem of information asymmetry.

This particular UIC was an individual performance for the professor, but the profit it generated was of the greatest gain for Mr. Miura. With their resulting profits, Pearl can subsidize colleges and universities. By now, Pearl has provided subsidies to Hiroshima University for 3 years, with JPY 3 million in the year before last, JPY 5 million in the last year, and JPY 2.5 million this year. The amount is closely related to the profit generated by UIC. Subsidies provided by the Hiroshima Industrial Promotion Organization do not have to be repaid: enterprises only need to pay tax if a profit is earned. Only when small- and medium-sized enterprises are supported, the governments can successfully collect taxes. The corporate tax of Japan is 30% for each year when profit exceeds JPY 8 million. Pearl maintains a cooperative relationship with Professor Watanabe. Meanwhile, it also cooperates with Kinki University in developing socks meant to promote maternal health and with the Prefectural University of Hiroshima in studies on socks preventing the elderly from mistaking the brake for the accelerator while driving. Mr. Miura believes that Pearl cannot exclusively rely on Hiroshima University, and that only wide cooperation with different universities can improve Pearl's overall level of performance. He also does not believe that enterprises should be limited to UIC, but rather that they should also perform university-university-industry collaboration to strive to establish additional collaborations between universities. As Sankei Shinbun [Japanese] newspaper reported, graduates in the Chugoku/Shikoku area are pleased with Hiroshima University, but not with the Prefectural University of Hiroshima. As the current economic situation of the Hiroshima Prefecture is at a depression, President Miura has attempted to cooperate with different universities and is doing his part to drive the economic development.

Annexure 3

Reference Case II: Cooperation in Product Development with the Graduate School of Science of Hiroshima University

After graduating from the University of Tokyo with a doctoral degree in 1993, Professor Ueno (上野) worked at the Institute for Molecular Science in Okazaki until 2005. At the age of twenty-eight, he became an associate professor at the Institute for Molecular Science, and eventually began his tenure as the head of a laboratory there. According to Professor Ueno himself, he was their youngest associate professor at that time. Six months later, after becoming a professor at the Institute for Molecular Science, Professor Ueno left for the Graduate School of Science of Hiroshima University to work on research and teaching.

Approximately 5 years ago, Professor Ueno began to engage in collaborative activities between universities and enterprises. Research funds were provided by JST, which holds various technology exhibitions for promotion in Tokyo and requires university professors to display their research achievements in their areas of expertise. Enterprises visit these exhibitions to find research results that match their demands and establish collaborative relationships with university professors.

Currently, enterprises are facing an environment in which new development is difficult, as they need a lot of new materials and new technologies. In the field of engineering, enterprises work very closely with universities' graduate schools of engineering science. Enterprises are clearly aware of the advanced technologies that universities have availed, a fact which makes it easy to cooperate with universities in research and development. However, in the field of science, enterprises have almost no information on the kinds of materials that universities can develop, or even how these materials might benefit them. Basic subjects are poorly known in the industry circle, and this is the biggest constraint to UIC in basic subjects.

Professor Ueno specializes in studies on organic molecular magnets, a field which attracted many enterprises to the idea of collaborating with him for different purposes, some for production of catalysts, some for manufacture of special plastics, and some for production of second-generation cells. The university conducted molecular synthesis in accordance with enterprises' requirements, and then submitted relevant data to them for product testing. After testing had been completed, the enterprises returned the results to the university for molecular design and modification. For example, in the catalyst collaboration, the Ueno Laboratory conducted molecular modification according to enterprises' feedback, and enterprises made catalysts out of modified molecules. Enterprises then sent relevant product data back to the university, and the Laboratory modified the molecules again for better effects. In addition, without testing the equipment and the reagents, enterprises cooperated with the university in order to be allowed to test their synthetic materials in the university's laboratory. In Professor Ueno's eyes, the results of his collaboration with enterprises were just unsatisfactory by-products. His studies are in-depth scientific research projects that cannot be put into practical use at present. There are twenty students in the Ueno Laboratory, and only one student participates in collaborative research and development with enterprises. However, although Professor Ueno was not very interested in this collaboration, the application of new materials in the collaboration boosted students' study, and students were increasingly interested in research after learning more about the practical uses of science. Moreover, Professor Ueno had the opportunity to observe the application of new materials, which widened the application of basic science in practical areas. Professor Ueno also held a post in the Cutting-edge Research Center, and researchers from various enterprises were sometimes employed as special researchers in the Center to participate in collaborative work. Engineers of enterprises gave lectures to students, and students' enthusiasm in scientific research was enhanced after witnessing the practical use of theories that they studied, an example being how one might apply TV materials to enterprises' products.

Professor Ueno did not value the research funds contributed by the collaboration with enterprises, but rather considered the collaborations a burden. The investment of these funds restricted his area of research and orientation. His research focused on the magnetic property of organic molecules, while enterprises were interested in research on the electrical property of organic molecules. Scientific research involving chemicals is considered basic, and a professor's research orientation can be changed according to an enterprise's demands. Therefore, Professor Ueno could collaborate with multiple enterprises simultaneously to develop various products that all related to the application research on organic molecules, but as enterprises' funds were accepted, his research projects would then be subject to inspecting by these enterprises. Professor Ueno is unwilling to collaborate with enterprises for that reason, preferring instead to apply to the government for scientific research funds to conduct research projects in which he has an active interest. As he himself put it, it was "take enterprises' money and serve them," which went against his belief of pursuing science. Though Professor Ueno did not like to be restricted by enterprises, he could be an active witness in the collaboration that resulted in the molecules he studied being presented in different forms; he also enjoyed the pleasures of this collaboration. For example, in the collaboration with four enterprises, the same organic molecules could be used to manufacture display materials, cells, color-changing glasses, rechargeable batteries, and even ink for printing. This is due to that the materials being dissoluble and transparent, meaning their color will change after charging: batteries dissolve in ink, and words printed out with this ink act as batteries. Ordinary batteries are made from metallic materials and graphite. Thus, many enterprises are seeking new, more efficient alternatives. This collaboration certainly also resulted in an increase in the Laboratory's number of theses and patents.

This kind of close UIC is not common in the Graduate School of Science. In the words of Li, a doctoral student of Professor Ueno, other laboratories in the School have few or almost no collaborations with enterprises. Li did not engage in the collaboration between Professor Ueno and enterprises. He seemed to have full of complaints about his own theoretical research, and was not very interested in it. However, when he talked about the collaboration with enterprises he knows very well, I clearly sensed much more enthusiasm.

Annexure 4

Reference Case III: Waseda University and Construction of the First Smart Town in Japan

The Honjo Smart Energy Town Project (本庄スマートエネルギータウン) refers to the construction of the first smart town in Japan. In May 2010, the Saitama Association of Corporate Executives proposed building an advanced ecological village in the Honjo-Waseda area to the governor of the Saitama Prefecture, a proposal made on the basis of a conception of regional management of Waseda University. The project was formally commenced in May 2011. The Honjo Campus was funded jointly by Waseda University, the Saitama Prefecture, and Honjo City. In 1982, Waseda University Honjo Senior High School was established. There are two graduate schools at the Honjo Campus, one of which is for Global Information and Telecommunication Studies (GITS) and the other for Environment and Energy. The two schools provided academic support for the construction of Honjo Smart Town.

Ota has acted as the manager of the UIGC Department of the project since 2010 and was involved in the coordination and management of the construction of Honjo Smart Town. Before that, Ota worked in WTLO for 2 years and then spent 10 years in the Incubation Center. About 12 years of experience in the promotion of UIC laid a solid foundation for his engagement in managing the project. However, the complexity and multiplicity of the Honjo Smart Town Project make Ota's job very challenging in spite of his abundant experience in departmental coordination and communication.

The Smart Town Project was organized and operated by the Committee of the Honjo-Waseda Project and the Operating Committee of the Honjo Smart Town Project (Figure A2). Both project groups consisted of members from the university, government, and enterprises. Members of the Committee of Project included: (i) a chairman (a professor of Waseda University), (ii) three vice-chairmen (the deputy mayor of Honjo City, a senior managing director of a consortium, and a professor of Waseda University), (iii) a professor of Waseda University (concurrently the operating chairman), (iv) URs (Urban Renaissance Agency) in the Saitama Prefecture and Honjo City, and (v) an Affair Bureau, consisting of the consortium, Honjo City, and the Urban Renaissance Agency. The Operating Committee consisted of an operating chairman, a vice operating chairman, university researchers and managing members, an affair bureau, and an expert consultant team. Therein, the vice operating chairman concurrently acted as the senior managing director of the consortium who worked for the vice-chairmen of the Project Committee, while

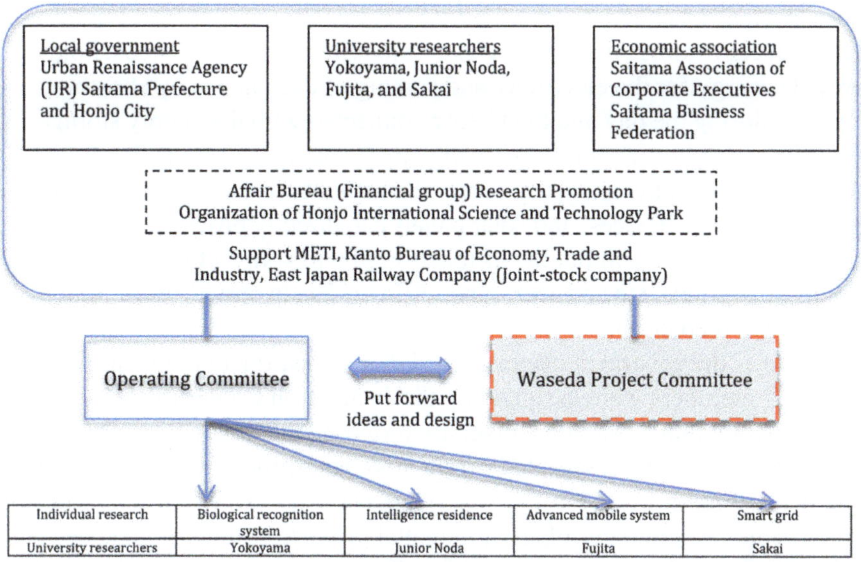

Figure A2 Organizational management structure of the Honjo smart town project.

the affair bureau was under the charge of the consortium, which also acted as the Affair Bureau of the Project Committee.

The entire project included four groups of researchers, which were a biological recognition mechanism group, a smart grid group, an advanced mobile systems group (e.g., electric cars, etc.), and a smart house group (Figure A2). The whole room, energy, and electric power were controlled by a microcomputer. The overall design of this area covered large shopping malls, Shinkansen stations, smart houses, and office buildings. It was estimated that 5,000 people would live in the town. Substitute fuel (biomass) was used as the fuel for the entire community. For example, alternative recyclable fuel (バイオマス) was used for electricity, heating, and automobile engine fuel. The goal of being economical and environmentally friendly was achieved using this substitute fuel. Since advanced devices were very expensive, the whole project was costly; complex and innovative technologies and equipment were funded by the government. The construction idea of the community was "sharing." A sharing network was formed for housing, shopping, and mobile tools, meant to operate individually or among households. Energy of the entire community came from the energy storage devices, but as these devices and solar panels are costly, it was decided that four households could share one such device for stored energy. Fortunately, electric cars themselves are essentially energy storage devices, and cars belonging to these four households can provide energy for the electricity and heating of each house by absorbing solar energy during the day and then being connected with the energy storage devices at night. If households save energy, they obtain eco-points (エコポイント), which can be used in the shopping center. Streetlights in the community are equipped with Wi-Fi devices, and solar panels can directly provide energy for cells and Wi-Fi devices without the installation of wires. In addition, the local government can connect to and access the Wi-Fi system to provide information like earthquake prediction.

Ota acted as the general manager of the UIGC Department of this project, and five staff members were assigned by the Incubation Center to engage in this project. As Ota said, Waseda University proposed the Smart Town Project to the government and then solicited cooperative enterprises from society via the Internet and the press. Many Japanese enterprises were interested in the Smart Town and technologies of environmental protection, and they had been searching for similar business opportunities. This project was exactly the right chance for them. To date, there have been more than fifty enterprises involved in this project.

Ota periodically participated in the joint meetings of the local government, university professors, and the industry association. Numerous organizations were involved in this project, so the project management was the most difficult matter for Ota. The local government was mainly in charge of approving the project and providing research funds, while the university provided support for technical research and patent permission. The economy association was responsible for communication and coordination related to the construction of enterprises. Today, the economic recession in Japan has caused financial strain on enterprises and slowed down the progress of construction. In addition, a shortage of manpower proved to be the most severe handicap for the project management. For this kind of multi-party cooperative project, coordination and management is the key for success. A significant portion of the funding provided by the government was used for research and development, so the coordination department in charge of Ota did not have enough money to employ people to do coordination and communication work. Additionally, effective information exchange was critical, and while university professors were always enthusiastic about creating new things, the government, which was more conservative by comparison, preferred to follow the existing pattern. These varying approaches made bilateral communication and exchange difficult. Frequently holding discussion meetings were the only way to resolve this conflict, and it was by this route that consensus was eventually reached between the university, government, and enterprises. One last key matter, in Ota's eyes, was that the three parties needed to settle upon a common goal. Ideally, the parties would work together to face and resolve difficulties encountered as they strove to achieve this goal, regardless of how difficult the cooperative process was.

Postscript

Those 6 years of graduate life in Beijing Normal University were fleeting, but they are engraved in my bones and printed on my heart. Beijing Normal University is a place where dreams begin and even more a place where dreams come true. With a yearning for the top university in the field of education, I entered this institution reverently. Three years later, the enthusiasm ignited by learning motivated me to stay at Beijing Normal University. From the moment I became a student there, Professor Yimin Gao, my teacher, has been an important guide in my life. Under his watch, I always believed that better days lie just ahead, regardless of whether I was walking through jungles or swamps in my academic career. He always entrusted me with important tasks with ease and made me unexpectedly discover another version of myself, one that was full of energy. His all-out support enabled me to seek self-growth in the days of traveling around Europe, and the colorful journey I made through Denmark, Belgium, the Czech Republic, France, and Sweden made me reinterpret the world. He did anything that he could to open a door for me, and enabled me to think about myself during my study in Japan so that I could be close to my original dream when I found that I was ready to return to it. Being sharpened by life and polished by Professor Gao, my sharpness and perseverance have become more mature and resilient.

Special thanks go to Professor Yutaka Otsuka of Hiroshima University for his fatherly love, encouragement, and support. With his help, I successfully completed my research in Japan. Mr. Otsuka helped me to contact and meet Mr. Matsui and later became a core contact person for all research. I am very grateful to Mr. Otsuka and Mr. Hosaka for introducing some important contact people for my interviews, and for later accompanying me during these interviews. They provided substantial support for ensuring that everything went smoothly. I would also like to thank the many Chinese students and Chinese people who helped me during my research in Hiroshima University. Their assistance facilitated my journeys to the University of Tokyo and Waseda University, thereby allowing me to successfully finish my research for my doctoral thesis. Here, I express my deepest appreciation to them. As the process of research was a long one, I knew the compiling of the thesis would be challenging and would require much endurance. Thanks go to Mr. Gao for his effort in providing instructions. He always led me confidently to the destination in my heart when I only had a vague outline of my idea in my head. With Mr. Gao's encouragement, trust, and support, I could, step by step, complete this doctoral paper in the way which I thought ideal.

Here in this study, I particularly express my sincere thanks to the China Scholarship Council, the Hong Kong Yuen Yuen Institute, and to Dr. Weiqi Tang, who provided free funding for the Joint School of Education, Social Science, and Medical Research Paper Award Program, as well as to Professor Cho Yee To, one initiator of the Joint School Paper Award Program, and all others who provided me the chance to perform practical investigations and studies in Japan.

Many thanks go to all teachers at the Institute of International and Comparative Education (IICE)! During 6 years of time there, the IICE has been my philosophical home, and every teacher there has acted as a sort of family member in my life. On my way forward in pursuit of my dreams, I will find the strength to persevere with your encouragement and support. You are the ones who have helped to make my dreams come true in my time here and I am sincerely grateful to you. I hope to create dreams and miracles of the future with your selfless love.

Thanks go to brothers and sisters of the University and public for their care and help. In having your company on the "bitter but sweet" way of study, my life is always a colorful landscape. Although it is not as profound as wash painting, it is flexible and yearns for the sun like Van Gogh's *Sunflowers*.

Thanks go to dearest father and mother for their long, most intense, and warmest love. You covered my weakness with warmth of the family and enabled me to be courageous and advance bravely.

Thanks go to all teachers and friends I met, became acquainted with, and knew well again! No matter how short or accidental our encounters were, you have built a part of my complete life; even though we may not meet each other again, you have been a power source in my advancement.

Thanks go to all people who love me and whom I love in return! In the end, I have written this for my grandfather in heaven. Thank you for your remote blessing and protection!

Acknowledgment

I wish to thank China Scholarship Council for funding my initial research in Japan, as well as many Japanese and Chinese students who helped me find my way in a strange land.

I owe a debt of gratitude to Professor Matsui Michikage—Coordinator of the International Industry–Academia Collaboration Division, the Center for Collaborative Research & Community Cooperation of UIC Center in Hiroshima University, who selflessly bore a heavy load of responsibility and concern in bringing this project to a successful end.

I'm particularly indebted to my supervisors—Professor Gao Yimin of the Institute of International and Comparative Education, Beijing Normal University, and Professor Yutaka Otsuka of Hiroshima University, whose guidance, encouragement, suggestion, and very constructive criticism have contributed immensely to the evolution of my ideas in this book.

It is also my duty to record my thankfulness to Professor Nishida Keiya of Hiroshima University, Professor Oshita Seiichi of the University of Tokyo, Professor Daisho Yasuhiro of Waseda University, Dr. Yamakawa Masahisa of the Mazda Motor Corporation, Amamiya Masahiro of the Zensho Company, Dr. Hirota Toshio as the visiting professor of Waseda University, Terasaki Takayuki of the Nissan Motor Co., Ltd. and Dr. Hosaka Yukio, all of whom have either inspired me or helped me undertake this project.

Finally, I take this opportunity to acknowledge Professor Du Xiangyun of Aalborg University, the whole publishing team, and everyone who collaborated in producing this book.

Zhiying Nian
March 2013

Matsui Michikage, in the middle of the first row, in UIC Center of Hiroshima University.

The graduation ceremony in Japan on Feb. 16, 2013.

Ph.D. defense in BNU.

With Dad and Mom.

My family with Professor Otsuka at home in Beijing in September, 2014.

Bibliography

[1] [GER] Wolfgang Kasper, Manfred E. Institutional Economics: Social Order and Public Policy [M]. Translated by Chaohua Han. Beijing: Commercial Press, 2000.

[2] [US] Ken Jones, Yanwen Zhou. University-Industry-Collaboration: Experience from the UK [J]. Exploring Education Development, 1991 (S3): [J]. Exploring Education Development, 1991 (4): 19–21.

[3] [US] Written by Burton R Clark, translated by Chengxu Wang et al. The Higher Education System: Academic Organization in Cross-National Perspective [M]. Hangzhou: Hangzhou University Press, 1994.

[4] [US] Edited by Lewis Branscomb, Richard Florida [JP] Fumio Kodama. Translated by Hongyi Yin, Jun Su. Industrializing Knowledge: University-Industry Linkages in Japan and the United States [M]. Beijing: Xinhua Publishing House, 2003.

[5] [US] Written by R. H. Coase, translated by Hong Sheng et al. The Firm, the Market, and the Law [M]. Shanghai: Shanghai Joint Publishing Company, 1990.

[6] [US] Written by D. C. North, translated by Yu Chen et al. Structure and Change in Economic History [M]. Shanghai: Shanghai Joint Publishing Company, 1994.

[7] [US] Written by D. C. North, translated by Shouying Liu. Institutions, Institutional Change and Economic Performance [M]. Shanghai: Shanghai Joint Publishing Company, 1994.

[8] [JP] Written by Ikuo Amano, translated by Wuyuan Chen. On Classification of Higher Education Institutions in Japan [J]. Fudan Education Forum, 2004 (5): 5–10.

[9] [JP] Masahiko Aoki. Toward a Comparative Institutional Analysis [M]. Translated by Li'an Zhou. Shanghai: Shanghai Far East Publishers, 2001: 28.

[10] [JP] Masahiko Aoki. Toward a Comparative Institutional Analysis: Motivations and Some Tentative Theorizing [A]. Translated by

Kuanping Sun. Transition, Regulation and System Selection [C]. Beijing: China Social Science Press, 2004: 129.

[11] Achievements of University-Industry-Collaboration between NEC Electronics (CN) and Beijing Institute of Technology Support Green Olympic [J]. Application of Electronic Technique, 2008 (9): 13.

[12] R. Coase et al. Property Rights and Institutional Changes [M]. Shanghai: Shanghai Joint Publishing Company, 1994.

[13] Correspondents of this Journal: Tsinghua University and TSMC Join Hands for A New Chapter of University-Industry-Collaboration [J]. Electronics & Packaging, 2010 (1): 46.

[14] Yong Cao, Shankun Liu. Japan's Research System for University-Industry-Collaboration [J]. Science Economy Society, 1994 (3): 27–29, 13.

[15] Changguo Chen et al. University-Industry-Collaboration: Exploration of a New Mode for Late-stage Undergraduate Education [J]. Higher Education in Chemical Engineering, 2000 (3): 45–48;

[16] Chunrong Chen. Improving the Quality of Dissertation Designing in Engineering Colleges through University-Industry-Collaboration Education Mode [J]. Higher Education Research in Mechanical Industry 2002 (S1): 26–28;

[17] Guanlong Chen et al. Develop University-Industry-Collaboration and Cultivate New-type Mechanical Engineering Talents [J]. China Higher Education Research, 2002 (22): 16–18.

[18] Guiyao Chen. A Study of the Mode for China's Universities to Participate in National Innovation System [D]. Doctoral Thesis of Zhejiang University, 2004: 201–234.

[19] Jiaoxuan Chen. Targeting the Market, Integration into the Regions and Emphasis on Capability - Shanghai University of Engineering Science: Innovation in University-Industry-Collaboration along with Talent Cultivating Mode [J]. Shanghai Education, 2008 (20): 30–31.

[20] Jing Chen, Xuewen Zhang. Collaboration Innovation of (IUG) Industry-University-Government Collaboration in Japan—Multiplex Angle of View of History-Pattern-Strategy and System [J]. Studies in Science of Science, 2008 (4): 880–886.

[21] Jing Chen. Study on the Knowledge Creation Process of University-Industry-Collaboration Innovation [D]. Zhejiang University Press, 2009.

[22] Jing Chen. Innovation and Development of University-Industry-Collaboration Strategic Alliances in the New Situation [M]. Beijing: China Renmin University Press, 2009.
[23] Edited by the Cihai Editorial Committee. Cihai Dictionary [M]. Shanghai Lexicographical Publishing House, 1980: 185.
[24] Chengqi Cui et al. Thinking and Practice of Application-oriented University in Strengthening Practice Teaching by Implementing University-Industry-Collaboration Education [J]. Journal of Guangdong Baiyun Institute, 2006 (1): 51–53, 57;
[25] Yingde Cui, Libin Cai et al. Exploration and Practice of University-Industry-Collaboration (Edition II) [M]. Zhongshan University Press, 2004.
[26] Yingde Cui, Libin Cai et al. Exploration and Practice of University-Industry-Collaboration [M]. Zhongshan University Press, 1999.
[27] Cunrui Deng. Measures for Teaching Reform in Higher Engineering Education in Developed Countries [J]. Overseas Higher Engineering Education, 1989 (1): 19–21.
[28] Hua Deng et al. University-Industry Interaction Mechanism in Chinese Higher Education Institution [J]. Tsinghua Journal of Education, 2006.
[29] Shujun Diao. The Rise and Development of Education Combining Learning and Production in the USA [J]. Journal of Wuyi University (Social Sciences Edition), 2004 (3): 92–95.
[30] Guowei Fang. Government's Role and Strategic Analysis in the Combination of Enterprise, School and Scientific Research [D]. Doctoral Thesis of Wuhan University, 2010.
[31] Pengju Gao et al. On the Research of Marketing through University-Industry-Collaboration – A Case Study of the University-Industry-Collaboration between Donghua University and Volkswagen Sales Company [J]. Exploring Education Development, 1999 (S3): 85–87.
[32] Jinlong Geng, Weiping Liu. Application and Inspirations of University-Industry-Collaboration in Japan's Vocational Education [J]. Career Horizon, 2007 (3): 51–53.
[33] Xiaojun Gong, Yinglong Ge, Yufang Chen. Study of the Long-term University-Industry-Collaboration Mechanism Based on Quality Management Mode [J]. Vocational Education Research, 2009 (10): 142–143.
[34] Guoqiang Gu et al. University-Industry-Collaboration, in a Supporting Role, Keeps One Step Ahead through Constant Innovation – How Does the Research Group for Hydraulic Pressure of Hydraulic Synchronizing

Integral Lifting of Tongji University Maintain Constant Major Scientific and Technological Tasks [J]. Research and Development Management, 1997 (2): 37–40.

[35] Jiafeng Gu. Research on the Management Mechanism of University-Industry-Collaboration – A Case Study of Beijing University [J]. Researches in Higher Education of Engineering, 2007 (3): 87–90.

[36] Yi Guo. University-Industry-Collaboration Talents with a Second Bachelor Degree – A Case Study of the Marketing Department of East China University of Science and Technology [J]. Exploring Education Development, 1999 (S3): 53–54.

[37] Jie He, Weili Fu, Yinghua Li. Inner Spirit and Application Prospect of Cooperative Education in China's Higher Education [J]. Liaoning Higher Education Research, 1995 (4): 11–15.

[38] Edited by Henry Etzkowitz et al, translated by Daoyuan Xia et al. Universities and the Global Knowledge Economy [M]. Nanchang: Jiangxi Education Publishing House, 2002.

[39] Enhua Hu, Xiuli Guo. The Existing Problems and Its Countermeasures Research of China's Production, Academic Studies and Research Institution Cooperation Innovation [J]. Scientific Management Research. 2002 (1): 69–72.

[40] Lizhi Huang. Inspirations of TAFE University-Industry-Collaboration in Australia for China's Higher Vocational Education [J]. Communication of Vocational Education, 2007 (3): 29–31.

[41] Shao'an Huang. Institutional Economics [M]. Beijing: China Higher Education Press, 2008.

[42] Xinchang Huang. Eight Measures of the Governments in Developed Countries for Promoting University-Industry-Collaboration Research [J]. Exploring Education Development, 1989 (4): 93–98.

[43] Xinchang Huang. American Universities - Main Drivers of Industrial Cooperation – NSF and Its University-Industry-Collaboration Programs [J]. Research and Development Management, 1990 (4): 73–74.

[44] Yiwu Huang. Comparison of Social Environment for Cooperative Education in China and America [J]. Journal of Yangtze University (Social Sciences Edition), 2005 (3): 127–129.

[45] Yingzhong Huang et al. Study of University-Industry-Collaboration Modes for Higher Management Education [J]. Journal of Chang Jung Christian University, 2002, 6 (2): 15–31.

[46] Hongdou Huo. Study of "University-Industry-Collaboration Education" in Institutions of Higher Education in the US [D]. Master Thesis of Liaoning Normal University, 2010.
[47] Yiping Jiang. Study of Development of Technical Human Resources through University-Industry-Collaboration Mode – Analysis on the Implementation of Industry-University Collaboration in Vocational Schools in Taiwan [J]. Journal of Technology, 2000, 15 (1): 139–148.
[48] Zili Kang. Cooperation Between Universities and Industries [M]. Taipei: Hanwen Publishing House, 1997: 47–63.
[49] Fuqi Li. University-Industry-Collaboration Cultivates High-quality Master of Engineering [J]. China Higher Education, 1993 (9): 15;
[50] Guishan Li, Yang Guo. University-Industry-Collaboration Education in Institutions of Higher Education in Canada and Its Reference Significance [J]. Social Sciences Abroad, 2010 (3): 108–113.
[51] Haizhi Li et al. University-Industry-Collaboration and Graduation Design (Thesis) of Engineering Majors [J]. Researches in Higher Education of Engineering, 2004 (1): 85–86.
[52] Meifang Li. Study of the Effectiveness of University-Industry-Collaboration [D]. Doctoral Thesis of Wuhan University of Technology, 2011.
[53] Wenhui Li, Chuhong Wang, Ning An. Analysis on the Subject-Object Relationship and Modes of University-Industry-Collaboration under the Innovation System [J]. Science and Technology Management Research, 2008 (6): 4–5.
[54] Xi Li. On the Development Mode of International Industrial R&D Hub [D]. Doctoral Thesis of East China Normal University, 2006.
[55] Xiaotao Li et al. A Comparative Study on University-Industry-Collaboration Education in Germany, Britain and Japan [J]. Journal of Wuhan Institute of Technology, 2008 (2): 117–120.
[56] Xinnan Li, Fuquan Sun. Towards the 21st Century: How Do Chinese Enterprises Step into the Knowledge Economy? [M]. Beijing: China Industry and Commerce Press, 1999.
[57] Zengping Li et al. Reform Practices of Talent Cultivation Mode through University-Industry-Collaboration in Botanical Garden Department [J]. Vocational and Technical Education in Xinjiang, 2004 (2): 45–48.
[58] Zhenxiang Li. Comparison of Internal Impetus for University-Industry-Collaboration in Higher Vocational Education between China and Foreign Countries [J]. Education and Vocation, 2010 (5): 156–157.

[59] Zhongxue Li. Practice and Theory of University-Industry-Collaboration in Cultivation of Master of Engineering [J]. Chinese Geological Education, 1999 (1): 52–55;
[60] Jianxin Li. University-Industry-Collaboration in Japan's Institutions of Higher Education and Inspirations [J]. Journal of Linyi Teachers' University, 2006 (1): 110–112.
[61] Yanhua Lian, Xiaoguang Ma. Development Trends Evaluation of Cooperation of Industry, University and Research Institute [J]. China Soft Science, 2001 (1): 54–59.
[62] Zongming Liao. Japan's "University-Industry-Collaboration" [J]. China Higher Education, 1994 (12): 38–39.
[63] Zongming Liao. On "University-Industry-Collaboration" of Institutions of Higher Education in Japan [J]. Tsinghua Journal of Education, 1994 (1): 117–122.
[64] Li Lin. Theoretical and Empirical Study of University-Enterprise Knowledge Alliance [M]. Science Press, 2010.
[65] Yifu Lin. On Institution and Institutional Change [J]. China: Development and Reform, 1988 (4).
[66] Guochen Liu et al. The Irresistible Trend of Reform of Institutions of Higher Education - University-Industry-Collaboration [J]. Liaoning Education Research, 1992 (5): 32–34.
[67] Li Liu. Historical and Comparative Study on University-Industry-Collaboration [D]. Doctoral Thesis of Zhejiang University, 2002.
[68] Ping Liu, Lian Zhang. Introduction to University-Industry-Collaboration Education [M]. Harbin Engineering University Press, 2007.
[69] Caichen Lu. Trend and Inspirations of Japan's University-Industry-Collaboration [J]. Journal of Liaoning Teachers College (Social Science Edition), 2006 (6): 100–102.
[70] Yongxiang Lu. Building up a National Innovation System Oriented to the Knowledge Economy Era. Guangming Daily, 1998/2/6.
[71] Yongxiang Lu et al. National Innovation System Oriented to the Knowledge Economy Era [M]. Beijing: Science Press, 2000.
[72] Wei Luo. Yuanhu Tang. The Reason and Motives for Firms to Participate in Cooperative Innovation [J]. Studies in Science of Science, 2001 (3): 65–68.
[73] Yan Luo, Ming Li. Study of the University-Industry-Collaboration Modes and Operation Mechanism in Local Universities and Colleges [M]. Sichuan Publishing Group Bashu Publishing House, 2009.

[74] Guimin Ma et al. Study of University-Industry-Collaboration Education Modes in Higher Vocational Schools [J]. Agriculture & Technology, 2010 (2): 125–127.

[75] Ning Ma. On the Innovation Modes of Enterprise-oriented University-Industry-Collaboration [J]. Studies in Science of Science, 2005 (12).

[76] Ning Ma. On the Allocation Modes of Scientific and Technological Resources in Enterprise-oriented University-Industry-Collaboration [J]. Research and Development Management, 2006 (5): 89–93.

[77] Weihua Ma. A Study on the Mechanism of the Effect of University-Industry Collaboration to the Core Competence of Academic Teams in Universities [D]. Doctoral Thesis of South China University of Technology, 2011.

[78] Yongbin Ma, Sunyu Wang. Exploration on Triple Helix Model of University-Government-Industry Partnership [J]. Researches in Higher Education of Engineering, 2008 (5): 28–34.

[79] Yongbin Ma. A Literature Review of University-Government-Enterprise Relationship [J]. Tsinghua Journal of Education, 2007 (5): 26–33.

[80] Changshun Nie. Base, Present Situation and Development Trend of Japan's University-Industry-Collaboration [J]. Japan Problem Studies, 1997 (1): 56–60.

[81] Yaofang Pan et al. The Present Situation of University-Industry-Collaboration in Higher Education and Exploration of Countermeasures [J]. Journal of Changchun University of Science and Technology (Higher Education Edition), 2010 (1): 75–77.

[82] Jianguo Qi et al. Technical Innovation - Reform And Restructuring of National Systems [M]. Economic Management Press, 1997.

[83] Lianqing Qiang. Inspirations from the Increase of Competition in Comprehensive National Strength between the US and Japan to Higher Education of China [J]. Fudan Education, 1995 (2): 6–11.

[84] Masahiko Aok, Harayama Yuko. Development Direction of University-Industry-Collaboration [J]. Technology Economics and Management, 2005 (4): 8–9.

[85] Xiuhua Ren, Xiaoqi Zheng, Hanbang Li. The Present Situation and Inspirations of China's University-Industry-Collaboration [J]. Heilongjiang Social Science, 2007 (2).

[86] Xiangqian Ruan. Effective Ways of Cultivating Innovative Master of Engineering [J]. Scientific and Technological Progress and Countermeasures, 2003 (16): 167–168;

[87] Huiwen Shang et al. Constructing Higher Vocational Talents Training Model about Instilling Study into Work by Industry-university Cooperation [J]. Education and Vocation, 2011 (3): 25–26.
[88] Enze Shen. Present Situation and Prospect of Japan's "University-Industry-Collaboration" [J]. Science of Science and Management of S&T, 1985 (3): 37–40.
[89] Xiaoqiu Shi. Exploration of University-Industry-Collaboration in Teaching Material Development for Application-oriented Undergraduate Programs [J]. China University Teaching, 2010 (2): 83–85.
[90] Fuquan Sun, Baoming Chen, Wenyan Wang. University-Industry-Collaboration Innovation in Major Developed Countries – Basic Experience and Inspirations [M]. Beijing: Economic Management Press, 2008.
[91] Shuchun Tan, On the Long-term University-Industry-Collaboration Mechanism in Higher Vocational Education [J]. Heilongjiang Researches on Higher Education, 2010 (7): 81–83.
[92] Yaowen Tang. Some Practices and Inspirations of University-Industry-Collaboration in Japan's Higher Vocational Schools [J]. Cultural and Educational Information, 2007 (16): 72–74.
[93] Hua Tian. A Study on the Development of Regional Universities Based on Knowledge Spillover [D]. Doctoral Thesis of Zhejiang University, 2010.
[94] Chengjun Wang, Peimin Wang. University-Industry-Collaboration or Government-University-Industry-Collaboration? [J]. Researches in Higher Education of Engineering, 2005 (1): 28–33.
[95] Chengjun Wang. A Study on the Government-Industry-University Partnership Triple Helix – Knowledge and Selection [M]. Social Sciences Academic Press, 2005.
[96] Chengyun Wang. On the Technological Innovation Mode and R&D Activities of Japanese Enterprises in China [D]. Doctoral Thesis of East China Normal University, 2008.
[97] Duanqing Wang. Higher Engineering Education Reform: from the University-Industry-Collaboration Perspective [J]. China Metallurgical Education, 1994 (1): 11–16;
[98] Fan Wang. The Present Situation of University-Industry-Collaboration in Japan's Universities [J]. World Education Information, 2007 (5): 60–64.

[99] Wenyan Wang, Fuquan Sun, Qiang Shen. Classification, Characteristics and Selection of University-Industry-Collaboration Modes [J]. Forum on Science and Technology in China, 2008 (5): 37–40.

[100] Yiming Wang, Jun Wang. Issues Concerning Improving Independent Innovation Ability of Enterprises [J]. China Soft Science, 2005 (7): 10–14.

[101] Yutong Wang et al. Effective Modes for Cultivating Application-oriented Postgraduates through University-Industry-Collaboration based on Education of Master Degree of Engineering [J]. Scientific Management Research, 2005 (3): 78–79, 86.

[102] Chunyan Wei, Lin Li. The Present Situation and Prospect of Japan's Higher Education Reform [J]. Studies in Foreign Education, 2000 (3): 41–45.

[103] Baosan Wu. Selections of History of Chinese Economic Thought (Pre-Qin Period) [M]. China Social Sciences Publishing House, 1996: 548.

[104] Hongyuan Wu, Xiaoqi Zheng. The "Coordinator" System and Inspirations in Japan's University-Industry-Collaboration [J]. Researches in Higher Education of Engineering, 2006 (3): 78–81.

[105] Yan Wu. New Concept, Target and Task: Thesis Collection of China's University-Industry-Collaboration Education Summit 2009 [M]. Beijing: Higher Education Press, 2010.

[106] Xiqi Xiao. On the Cooperative Education in China's Industrial Education [J]. Journal of Education Information, 1997 (19): 161–144.

[107] Haifeng Xu, Defang Li. Measures and Inspirations of Japan's University-Industry-Collaboration under New Circumstances [J]. Career and Adult Education, 2006 (11): 40–41.

[108] Hui Xu. A New Stage of Development of Higher Education: On the University and Industrial Relations [M]. Hangzhou: Hangzhou University Press, 1990: 1.

[109] Changqing Xu. International Survey on the New University-Industry-Collaboration: Case Study Based on Japan [J]. Higher Education Exploration, 2008 (5): 60–65.

[110] Huiying Xu. Experiences and Lessons from the UK's University-Industry-Collaboration [J]. Science & Technology Industry of China, 2010 (11): 70–72.

[111] Lang Xu, Dong Liu. Build up an Industry-University-Research Institution Joint Innovation Mechanism Dominated by Enterprises [J]. Modern Enterprise. 2006 (11): 14–15.

[112] Xingmiao Xu, Baiyi Wang. Innovation in Industry-University-Research Institute Collaboration in Finland [J]. Education Review, 2011 (3): 156–158.

[113] Yingmei Xue, Yingzhi Zhou. Study of Cooperative Models between Industry and Universities or Scientific Institutes and Relevant Issues in China [J]. Journal of Shandong Medical University (Social Sciences Edition), 2000 (2).

[114] Qing Yan, Qing Xu, Zhangzhong Wang. Study on Selection of Independent Innovation Mode Based on Production-Education-Research Institution Collaboration [J]. Higher Education Exploration, 2008 (1): 65–68.

[115] Xinqun Yang et al. General Motivation for Combination of Industry, University and Research Institution-A Case Study of the Cooperation Between Beijing University of Chinese Medicine and Wantong Pharmacy Group [J]. Journal of the Knowledge Economy, 2010 (18): 7.

[116] Xiufen Yang. A Study of the University-Industry Cooperation Model and Innovative Performance in Taiwan [D]. Doctoral Thesis of Jilin University, 2010.

[117] Yan Yang. Research on Cooperation Mode of Production-Education-Research Institution [J]. Sci-Tech Information Development & Economy, 2010, 20 (7): 179–180.

[118] Wei Yao, Jing Chen. Study on the Knowledge Creation Process of University-Industry-Research Institution Collaboration [M]. Hangzhou: Zhejiang University Press, 2010.

[119] Baifang Yu. [J]. University-Industry-Collaboration is an Effective Way to Achieve Objectives of Talent Cultivation [J]. Exploring Education Development, 1992 (3): 42–44, 27.

[120] Qingming Yuan. New Institutional Economics [M]. Beijing: China Development Press, 2005.

[121] Hui Zhang, Wanmin Wu. On the Long-term Mechanism of University-Industry-Collaboration in Higher Vocational Education [J]. Journal of Higher Education, 2008 (11): 67–72.

[122] Lihua Zhang. Investigation on Scientific Research System of Industry-University-Research Institution Collaboration in Japan [J]. Science and Technology Policy and Development Strategy, 2000 (6).

[123] Lian Zhang. Theoretical Problems in Industry-University-Research Institution Cooperative Education and Practices in China [J]. Vocational and Technical Education, 2002 (34): 21–25.

[124] Qiying Zhang et al. South Korea's Higher Vocational Education in University-Industry-Collaboration Mode [J]. Vocational Education Research, 2009 (7): 155–156.

[125] Shuyi Zhang. "University-Industry-Collaboration" Modes of Higher Vocational Education Abroad and Inspirations [J]. Xinyang Normal University (Philosophy and Social Science Edition), 2009 (7): 73–76.

[126] Wei Zhang et al. Strengthen Industry-University Cooperation and Open up New Ways of Cultivating High-level Talents [J]. Academic Degrees & Graduate Education, 1992 (5): 56–58.

[127] Wei Zhang. Research on the IUR Cooperation Behavior and Micro-Mechanism under the Regional System of Innovation [D]. Doctoral Thesis of Wuhan University of Technology, 2009.

[128] Wuchang Zhang. Economic Interpretation [M]. Beijing: Commercial Press, 2000: 455.

[129] Ximei Zhang. Comparison and Inspirations of Industry-University Cooperative Education in Developed Countries [J]. Research and Development Management, 1994 (2): 76–80;

[130] Xuejun Zhang. Learn the Scientific Outlook on Development in an In-depth Way Unremittingly and Unwaveringly and Devote Every Effort to Assist Employment of Regular University Graduates in 2010 [J]. China University Students Career Guide, 2009 (21): 4–9.

[131] Xuewen Zhang. Knowledge-based University-Industry-Collaboration Innovation: Boundary and Path [D]. Doctoral Thesis of Zhejiang University, 2010.

[132] Yong Zhang. Integrating Industry, University and Research Institution under NIS: Theory Discussion and Demonstration Research [D]. Master Thesis of Ocean University of China, 2006.

[133] Yu Zhang. Exploration and Practice on Innovative Talent Training Mode in Industry-University Cooperative Education [J]. Computer Education, 2010 (7): 12–14.

[134] Hanqiang Zhao et al. Inspirations of Overseas Cooperative Education of Production and Learning for China's Implementation of the Plan of Cultivating Excellent Engineers [J]. Higher Education of Sciences, 2010 (4): 49–52.

[135] Xuhui Zheng, Songqing Liu. Comparison between China and Foreign Countries in Motivations of Industry-University Cooperative Education [J]. Researches in Higher Education of Engineering, 2004 (3): 27–30.

[136] Ruizhi Zhi: A Research on Japan University Spin-off Companies in Regional Innovation Perspective [D]. Doctoral Thesis of East China Normal University, 2007.

[137] Education Division of Consulate-General of the People's Republic of China in Asaka. Japan Prepares to Implement Evaluation Measures in Universities in Regard to "University-Industry-Collaboration" [J]. World Education Information, 1999 (10): 27.

[138] Binglin Zhong. Carry out Industry-University-Research Cooperative Education to Cultivate Highly Competent Talent with Innovative Mind and Practical Abilities [J]. China Higher Education, 2000 (21): 15–17.

[139] Jian Zhong. The Institutional Analysis on the Silicon Valley Models Around the World [M]. Beijing: China Social Sciences Publishing House, 2000.

[140] Danong Zhou et al. Primary Vocational Talent Training Mode: Integration of Vocational Qualification Certificate and Academic Certificate, University-Industry-Collaboration [J]. Changzhou Institute of Light Industry Technology, 2006 (2): 7–10.

[141] Xiaofeng Zhou. Inspirations of Industry-University Cooperative Education in US Universities for China's Vocational Education [J]. Journal of Northwest Vocational Education, 2007 (10): 12–13.

[142] Zhiqiang Zhou, Yuming Yuan, Hongbo Gu, Shixin Li. Game Theory Analysis on Lack of Motivation in University-Industry-Collaboration in Higher Education. Contemporary Education Theory and Practice, 2010 (2): 37–40.

[143] Zhiqiang Zhou et al. Study of Lack of Motivation in University-Industry-Collaboration in Higher Education [J]. Contemporary Education Theory and Practice, 2010 (4): 60–62.

[144] Zhiqiang Zhou et al. Game Theory Analysis on Lack of Motivation in University-Industry-Collaboration in Higher Education [J]. Contemporary Education Theory and Practice, 2010 (1): 37–40.

[145] Guoren Zhu. Transition from "Developing the Nation via Science and Technology" to "Developing the Nation via Scientific and Technological Innovation"-Japanese Higher Education Meets the Challenge of Knowledge Economy [J]. Journal of Higher Education, 2000, (04).

[146] Jianyu Zhu. Inspiration from the "University-Industry-Collaboration" Center of University West [J]. Global Education, 2008 (11): 52–54;

[147] Jinlan Zhu. A Comparative Study on German Dual System and Japanese Production and Education Cooperation [J]. Journal of Jiangsu Teachers University of Technology, 2004 (3): 44–50.

[148] Zibin Zhu et al. The Inevitable Road to Higher Engineering Education-University-Industry Collaboration [J]. China Higher Education Research, 1998 (3): 48.

[149] Zibin Zhu et al. Study of University-Industry-Collaboration Mode in Higher School of Engineering [J]. Exploring Education Development, 1999 (S3): 38–41, 73.

[150] Adam, B. J. *The U.S. Patent System in Transition: Policy Innovation and the Innovation Process* [C]. Nber Working Paper Series, 1999.

[151] Adams, James D., Chiang. Eric P. and Starkey, Katara. Industry-University Cooperative Research Centers [J]. *Journal of Technology Transfer*, 2001, 26 (1–2): 73–86.

[152] Allen, Kathleen R. and Taylor, Cyrus C. Bringing Engineering Research to Market: How Universities, Industry and Government are Attempting to Solve the Problem [J]. *Engineering Management Journal*, 2005, 17 (3): 42–48.

[153] Annamária Inzelt. The evolution of university-industry-government relationships during transition [J]. *Research Policy*, 2004 (33): 975–995.

[154] Aokimasahi, H. Y. Industry-University Cooperation to Take on Here from. Research Institute of Economy [J]. *Trade and Industry*, 2002 (4): 42–49.

[155] Bauman Zygmunt. *Liquid times: living in an age of uncertainty* [M]. Cambridge: Polity Press, 2007.

[156] Bell, D. *The Coming of Post-industrial Society: A Venture in Social Forecasting*. New York: Basic Books, 1973.

[157] Bloedon, R.V. & Stokes, D. R. Making University Industry Collaborative Research Succeed Research [J]. *Technology Management*, 1994, 37 (2): 44–49.

[158] Bonaccorsi, A. & Piccaluga, A. A theoretical framework for the evaluation of university-industry relationships [J]. *R&D Management*, 1994, 24 (3): 229–247.

[159] Borys, B. & Jemison, D. Hybrid arrangements as strategic alliances: theoretical issues and organizational combinations [J]. *Academy of Management Review*, 1989, 14: 77–85.

[160] Caloghirou, Y. T. University-industry cooperation in the context of the European Framework Programmes [J]. *Journal of Technology Transfer,* 2001, *26* (1–2), 153–161.

[161] Carayannis, E.G., Popescu, D., Sipp, C. & Stewart, M.D. Technology learning for entrepreneurial development in the knowledge economy

(KE): Case studies and lessons learned [J]. *Technovation*, 2006, 26: 419–443.

[162] Carnegie Commission on Higher Education. *A Classification of Higher Education* [A]. 1976.

[163] Cohen, W. F. Industry and the academy: uneasy partners in the cause of technological advance. In R. Noll, *Challenges to the University.* Washington DC: Brookings Institution Press, 1998.

[164] Cohen, W. M., Nelson, R. R. & Walsh, J. P. Links and Impacts: The Influence of Public Research on Industrial R&D [J]. *Management Science*, 2002, 48 (1): 1–23.

[165] Corsten, H. Technology transfer from universities to small and medium-sized enterprises – an empirical survey from the standpoint of such enterprise [J]. *Technovation*, 1987, 6: 57–68.

[166] Dorf, R.C. & Worthington, K.K.F. Technology transfer from universities and research laboratories [J]. *Technology Forecasting and Social Change*, 1990, 39: 251–266.

[167] Drejer, I. & Jørgensen, B. H. The dynamic creation of knowledge: Analyzing public-private collaboration [J]. *Technovation*, 2005 (25): 83–94.

[168] Drucker P. *The Age of Discontinuity: Guidelines to Our Changing Society*. New York: MIT Press, 1968.

[169] Etzkowitz, H. and Leydesdorff, L. The Dynamics of Innovation: From National Systems and 'Mode 2' to a Triple Helix of University-Industry-Government Relations [J]. *Research Policy*, 2000 (29): 109–123.

[170] Etzkowitz, Henry, Asplund, Patrik, and Nordman, Niklas. Beyond Humboldt: the Entrepreneurial University, the Third Mission and the Triple Helix [J]. *VEST Journal for Science and Technology Studies*, 2003, 16 (1): 21–45.

[171] Freeman C. *Technology and Economic Performance: Lessons from Japan* [M]. Printer Publish, 1987.

[172] Geisler, E. & Rubenstein, A. *University-Industry Relations: A Review of 82*, 1989.

[173] Geisler, E. Industry-university technology cooperation: a theory of inter-organizational relationships [J]. *Technology Analysis & Strategic Management*, 1995, 7 (2): 217–229.

[174] Goldfarb, B. a. Bottom-up versus top-down policies towards the commercialization of university intellectual property. *Research Policy*, 2003, 32 (4): 639–658.

[175] Granovetter, M. E. Types of Knowledge and Their Roles in Technology Transfer [J]. *Journal of Technology Transfer*, 1985, 91 (3): 481–510.

[176] Gregory Mike. The Collaboration Challenge [J]. *Manufacturing Engineering*, 2005, 83 (6): 6.

[177] Gulbrandsen Magnus & Smeby Jens-Christian. Industry Funding and University Professors' Research Performance [J]. *Research Policy*, 2005, 34 (6): 932–950.

[178] Hall, B. H. Market value and patent citations. *RAND Journal of Economics*, 2005 (36): 16–38.

[179] Hall, B. L. *Universities as research partners.* NBER Working Papers N 7643, 2000.

[180] Henderson, R. J. Universities as a source of commercial technology: A detailed analysis of university patenting, 1965–1988 [J]. *The Review of Economics and Statistics*, 1998 (80): 119–127.

[181] Henry Etzkowitz & Loet Leydesdorff. The Dynamics of Innovation: From National Systems and 'Model 2' to a Triple University-Industry-Governmental Relation [J]. *Research Policy*, 2000 (29): 109–123.

[182] Henry Etzkowitz & Loet Leydesdorff. The Dynamics of Innovation: From National Systems and 'Model 2' to a Triple University-Industry-Governmental Relation [J]. *Research Policy*, 2000 (29): 109–123.

[183] Henry Etzkowitz. *The evolution of entrepreneurial university.* Int, J. Technology and Globalization, Vol. 1, No. 1, 2004.

[184] Kanama, D. a. A study on university patent portfolios (II): the impact of intellectual property related policies and the change into corporation of national university on patent application by national university corporation, 2008.

[185] Lanjouw, J. O. Patente quality and research productivity: Measuring innovation with multiple indicators. *The Economic Journal,* 2004 (114): 441–465.

[186] Laveand Wenger, E.J. *Situated Learning: legitimate peripheral participation*. Cambridge: Cambridge University Press, 1991.

[187] Leydesdorff, L. a. The decline of university patenting and the end of the Bayh-Dole effect [J]. *Scientometrics*, 2010 (83): 355–362.

[188] Link, A. Research joint ventures: evidence from fedral register filings. *Review of Industrial Organization,* 1996 (11): 617–628.

[189] Lissoni, F. L. Academic patenting in Europe: new evidence from KEINS database. *Research Evaluation*, 2008, 17 (2): 87–102.

[190] Liu, H. & Jiang, Y. Technology transfer from higher education institutions to industry in China: nature and implications [J]. *Technovation*, 2001, 21: 175–188.

[191] Liyanage, Shantha. Breeding innovation clusters through collaborative research networks [J]. *Technovation*, 1995 (9): 553–567.

[192] Loet Leydesdorff & Martin Meyer. Triple Helix indicators of knowledge-based innovation systems [J]. *Research Policy*, Vol. 35, 2006: 154–176.

[193] Loet Leydesdorff. The Triple Helix Model and the Study of Knowledge-Based Innovation Systems [J]. *Int. Journal of Contemporary Sociology*, 2005, 42 (1): 12–27.

[194] Low, M. *Japan: from technology to science policy*. Etzkowttez, H. and Leydesdorff, L. (ed.). University and the Globe Economy, 1997: 132–140.

[195] Matkin, G.W. *Technology Transfer and the University* [M]. Macmillan, New York, 1990.

[196] Michael, G., Linoges, C., Nowotny, H. Schwartzman, S., Scott, P. and Trow, M. *The New Production of Knowledge: the Dynamics of Science and Research in Contemporary Societies* [M]. London Sage Publications Ltd, 1994.

[197] Motohashi Kazuyuki. University-Industry Collaboration in Japan: The Role of New Technology-based Firms in Transforming the National Innovation System [J]. *Research Policy*, 2005, 34 (5): 583–594.

[198] Mowery, D. C. The Bayh-Dole act of 1980 and university-industry technology transfer: A Model for other OECD governments? [J]. *Journal of Technology Transfer*, 2005, 40 (1/2), 115–127.

[199] Mowery, D. D. Academic Patent Quality and Quantity before and after the Bayh-Dole Act in the United States [J]. *Research Policy*, 2002 (31): 399–418.

[200] Parker, D.D. & Zilberman, D. University technology transfers: impacts on local and US economies [J]. *Contemporary Policy*, Issues 11, 1993 (2): 87–96.

[201] Peter & Fusfeld, H. *University industry research relationships* [M]. National Science Foundation, USA, 1982.

[202] Ruth, S.K. Successful business alliance, classroom strategies [J]. *The Methodology of business education*. 1996 (34): 10–23.

[203] S. Siegel Donald, A. Waldman David, E. Atwater Leanne, N. Link Albert. Commercial Knowledge Transfers from Universities to Firms: Improving the Effectiveness of University-Industry Collaboration [J].

Journal of High Technology Management Research, 2003, 14 (1): 111–134.

[204] Sampat, B. N. Changes in university patent quality after the Bayh-Dole act: A re-examination. *International Journal of Industrial Organization*, 2003 (21): 1371–1390.

[205] Santoro, M.D. & Chakrabarti, A.K. Firm size and technology centrality in industry-university interactions [J]. *Research Policy*, 2002 (31): 1163–1180.

[206] Senker. A. Rationale for Partnerships: building national innovation systems [J]. *STI Review*. 1998 (23): 23–37.

[207] Tamada, S. a. Analysis of patents jointly applied for by universities or public research organizations and private sector firms [J]. *RIETI Discussion Paper Series*, 2007, 08-J-003.

[208] Thursby, J. G. Who is selling the ivory tower? Sources of growth in university licensing [J]. *Management Science*, 2002 (48): 90–104.

[209] Toffler A. (1990). Powershift: Knowledge, Wealth and Violence at the Edge of 21st Century. New York: Bantam Books.

[210] Tornatzky, Louis G. and Bauman, Joel S. *Outlaws or heroes?: issues of faculty rewards, organizational culture, and university-industry technology transfer* [M]. Southern Technology Council, 1997.

[211] Tornatzky, Louis, Waugaman, Paul and Denis Gray. *Innovation U: New University Roles in a Knowledge Economy* [M]. RTP, NC: Southern Growth Policies Board, 2002.

[212] Trott, P., Cordey-Hayes, M. & Seaton, R.A.F. Inward technology transfer as an interactive process [J]. *Technovation*, 1995, 15 (1): 25–43.

[213] United Nations Industrial Development Organization. *Structural Change in the World Economy: Main Features and Trends* [R]. Working paper 24/2009, Research and Statistics Branck, Vienna, 2010.

[214] WEF. *The Global Competitiveness Report 2010–2011.*

[215] WIPO. *Technology transfer, intellectual property and effective university-industry partnerships: the experience of China, India, Japan, Philippines, the Republic of Korea, Singapore and Thailand* [R]. WIPO, 2007.

[216] World Economic Forum. *The Global Competitiveness Report 2010–2011* [R]. World Economic Forum, 2010.

[217] Yong S.Lee. The Sustainability of University-Industry Research Collaboration: An Empirical [J]. *Journal of Technology Transfer*, 2000, 25 (2): 111–112.

[218] イノベーション25戦略会議. 長期戦略指針"イノベーション 25": 未来をつくる, 無限可能性への挑戦 [R]. 閣議決定, 2007, 6.
[219] 伊藤昭男. 産学連携と地域イノベーション [J]. 北見大学論集, 2000, (23): 13–35.
[220] 磯谷桂介. 日本の産学連携と大学改革の進展 [J]. 経済産業ジャーノル, 2004, 5.
[221] 科学技術•学術審議会技術•研究基盤部会. 大学等における産学官連携の現状について [R]. 産学官連携推進委員会（第５期第１回）, 2009.
[222] 科学技術政策研究所, 株三菱総合研究所. 主要な産学官連携? イノベーション振興の達成効果及び問題点 [R]. NISTEP Report No. 87, 2005. 3.
[223] 技術革新システム小委員会. 産学官連携の促進に向けて [R]. 2001,11.
[224] 近藤正幸. 科学技術における日本の政策革新–科学技術政策からイノベーション政策へ [J]. 研究技術計画, 2004, (19): 132–140.
[225] 近藤正幸. 大学発ベンチャーの育成戦略 [R]. 中央経済社, 2002.
[226] 近藤正幸. 日本の大学発ベンチャーの産業別? 地域別? 起業者別特性 [J]. 研究技術計画, 2005, (20): 90–102.
[227] 元桥一之. 産学連携の実態と研究開発型中小企業の重要性 [J]. 経済産業ジャーノル, 2004, 5.
[228] 元桥一之. 産学連携の実態と効果に関する計量分析: 日本のイノベーション改革に対するインプリケーション [R]. RIETI Discussion Papers Series 2003-J-015.
[229] 原山優子. 産学連携:「革新力」を高める制度設計に向けて [M]. 東洋経済, 2003.
[230] 坂元耕三, 近藤正幸. 産学共同研究に関する時系列分析及び企業特性別分析 [J]. 開発技術, 2004c, (10): 11–26.
[231] 三井逸友. 地域イノベーションシステムと地域経済復活の道 [J]. 信金中金月報, 信金中央金庫, 2004, (13): 2–25.
[232] 産学官連携推進委員会. 新時代の産学官連携の構築に向けて [R]. 文部科学省, 2003.
[233] 新堀通也. 日本の大學教授市場 [M]. 東京: 東洋館出版社, 昭和四〇年(1965年).
[234] 西村吉雄. 産学連携—中央研究所の時代を超えて [M]. 日経 BP, 2003.
[235] 斉藤明. 産学協同と開かれた大学 [J]. 専修商学論集, 1984, (37): 1–76.
[236] 総合科学技術会議. 第４期科学技術基本計画 [R]. 2011 年 8 月 19 日: 48.
[237] 中教審の答申については. 現代教育研究所. 中教審と教育改革 [M]. 東京: 三一書房, 昭和四六年.
[238] 内閣府. 科学技術政策の論点一科学技術政策の進捗状況と今後の課題一[R]. 社団法人時事画報社, 2004.
[239] 馬場, 後藤. 産学連携の実証研究 [M]. 東京: 東京大学出版会, 2007.
[240] 尾身幸次. 科学技術立国論 科学技術基本法解説 [N]. 読売新聞社, 1996.
[241] 澤田芳郎. 現代社会における科学と産業産学協同論のフレームワーク [R]. 京都大学教育学部紀要, 1990, (36): 163–184.

[242] 澤田芳郎. 社会組織イノベーションとしての産学連携—産学連携の三層モデルの視点 [R]. 産学連携学会第１回大会講演予稿集, 2003 (1): 63–64.

[243] Jinfu Bai. Diagnose the Drawbacks in Large State-owned Enterprise Innovation is Crucial to Promotion of Future National Competitiveness. News Weekend, 2007 (45). http://203.192.6.66/htm/content_1397.htm. [2011-12-02]

[244] Xin Fang. What is the National System of Innovation. http://www.oean.com.cn/ln/chuangxin.htm. 2002, 2, 18.

[245] people.com.cn. President Hu Jintao Emphasizes: Give Full Play to the Significant Role of Scientific and Technological Progress and Innovation [EB/OL]. http://www.people.com.cn/GB/shizheng/1024/3085597.html. [2011-3-6]

[246] Qingshan Wu, Tianyou Lin. University-Industry-Collaboration [J]. Educational Resources and Research, 2004: 59–115. Website: http://www.etf.europa.eu/web.nsf/pages/home. [2011-11-26]

[247] China Industry-University-Research Institute Collaboration Association. Five Problems in Cooperative Innovation of Industry-University-Research Institute. 2010-01-22 http://www.360cxy.cn/front/InfoTemp.aspx?InfoID=667008. [2011-12-02]

[248] OECD. National Innovation Systems [R]. OECD Publications, France, 1997. http://www.oecd.org/dataoecd/35/56/2101733.pdf. [2012-1-5]

[249] 文部科学省. 産学官共創力の強化への取組, 平成２２年１２月２１日 http://www.kantei.go.jp/jp/singi/titeki2/tyousakai/kyousouryoku/2011dai3/siryou3_1.pdf#search='大学等における産学連携' [2011-11-12]

[250] 長野裕子. 文部科学省科学技術政策研究所. 日本の大学等における産学連携の実態と意識動向, [2010-1-27] http://www.tsukuba-network.jp/sangakukan/pdf/material04.pdf#search='大学等における産学連携' [2011-11-12]

[251] 文部科学省. 平成18年度大学等における産学連携等実施状況について http:// www.mext.go.jp/a_menu/shinkou/sangaku/sangakub/07083106.htm [2011-11-12]

[252] 文部科学省. 平成21年度大学等における産学連携等実施状況について http://www.mext.go.jp/a_menu/shinkou/sangaku/1296577.htm [2011-11-12]

[253] Ministry of Education, Culture, Sports, Science, and Technology, Research Promotion Bureau. Committee for Evaluation of Supports to University-Industry-Government Collaboration: Evaluation Report of Supports to University-Industry-Government Collaboration [R].

March Heisei 17 (2005), http://www.mext.go.jp/b-menu/houdou/17/04/05042801/003.pdf. [2011-12-20]
[254] 科学技術・学術政策局産業連携・地域支援課大学技術移転推進室. 平成 22 年度大学等における産学連携等実施状況について [R]. 平成 23 年 11 月 30 日 (2011年) http://www.mext.go.jp/a_menu/shinkou/sangaku/__icsFiles/afieldfile/2011/11/30/b1313463_01.pdf. [2011-12-20]
[255] Science and Technology Basic Plan. (1996). Date of retrieval: October 28, 2012, source: MEXT: www.mext.go.jp/b_menu/shingi/kagaku/kihonkei/honbun.htm. [2012-12-20]
[256] 技術革新を目指す科学技術政策ー新産業創造に向けた産業技術戦略ー. (2005). Date of retrieval: November 26, 2012, Source: MEXT http://www.mext.go.jp/report/data/g50223aj.html. [2012-12-26]

Index

About the Author

Nian Zhiying (1978–), is the education advisor to the President of NetDragon (Fujian) Computer Network Information Technology Co Ltd in Fuzhou, Fujian Province. She is also a researcher from Beijing Institute for the Learning Society, Beijing Normal University.

She got her M.A. and Ph.D. degree in comparative education in the Faculty of Education at Beijing Normal University in 2010 and 2013, and then finished the postdoctoral research in School of Government in Beijing Normal University within two years. Her research field covers lifelong learning, comparative higher education and entrepreneurship education. She is working on a number of research projects on lifelong learning collaborated with Beijing Municipal Government, and some international and national research projects as well. She has involved in two edited books and three translated books published in China, such as *The Introduction to Problem-Based-Learning from the International Perspective: Theory and Practice* by Higher Education Press. She has published over 10 articles in Chinese journals.

E-mail: zhiynian_bnu@126.com

For Product Safety Concerns and Information please contact our EU
representative GPSR@taylorandfrancis.com
Taylor & Francis Verlag GmbH, Kaufingerstraße 24, 80331 München, Germany

www.ingramcontent.com/pod-product-compliance
Ingram Content Group UK Ltd.
Pitfield, Milton Keynes, MK11 3LW, UK
UKHW022210190926
475702UK00003B/294

* 9 7 8 8 7 9 3 3 7 9 0 4 6 *